AIRCRAFT MONOGRAPH

JANUSZ LEDWOCH

MESSERSCHMITT Bf 110

AJ•PRESS

AIRCRAFT MONOGRAPH 3

Front cover: The Messerschmitt Bf 110C–2 of I>/ZG 76 in combat with Hurricane fighters over Southern England, late 1940. *Paint:* *Jarosław Wróbel*

P. O. Box 73
80–461 GDAŃSK 45
tel. (0-58) 56-04-23, 56-34-76

English Edition:
BOOKS INTERNATIONAL
69B Lynchford Road
Farnborough, Hampshire GU14 6EJ, England
Telephone: 0252–376564, Fax 0252–370181

Editor in Chief: **Adam Jarski**
Cover painting: **Jarosław Wróbel**
Colour Plates: **J. Wróbel, S. Zajączkowski**
Book Design: **Jarosław Wróbel, Adam Jarski**
Assistant Editor: **Krzysztof Machała**
Drawings: **Krzysztof M. Żurek, Witold Hazuka, Archiwum**
English Edition edited by: **Richard Ward** and **Krzysztof Machała**

Printed in Poland by: MATRIX
Pruszków, ul. Focha 4

ISBN 83–86208–12–0

BOOKS INTERNATIONAL TRADE REPRESENTATION IN EUROPE:

POLAND, CZECH REPUBLIC, SLOVAKIA, HUNGARY, RUSSIA AND EASTERN EUROPE:
BOOKS INTERNATIONAL
Ul. Lubelska 30/32
03–308 Warszawa
POLAND
Tel. & Fax (2) 619–60–57

AUSTRIA, SWITZERLAND, ITALY:
Mr. Juliusz Komarnicki & Mr. Patrick Bygate
Casella postale 196
CH–6900 Massagno
Switzerland
Tel.: 41-91-571539. Fax: 41-91-567865

BELGIUM, LUXEMBOURG, The NETHERLANDS
Mr. Reinier Pleysier
PO Box 166
8180 AD Heerde
Netherlands
Tel.: 31-5782-5203. Fax: 31 5782 5243

GERMANY
Mr. Robbert Pleysier
Walkottelanden 72
7542 MT Enschede
Netherlands
Tel.: 31-53775377. Fax: 31 53 778298

FRANCE, GREECE, FORMER YUGOSLAVIA
Mr. Graham Powell
11 Meadow Court Close
Bournemouth, Dorset BH9 2BT
England
Tel.: 0202 528263. Fax: 0202 537595

PORTUGAL, SPAIN, GIBRALTAR
Mr. Patrick Bygate
Casella postale 196
CH–6900 Massagno
Switzerland
Tel.: 41-91-571539. Fax: 41-91-567865

DENMARK, FINLAND, ICELAND, NORWAY, SWEDEN
Mr. Juliusz Komarnicki
Casella postale 196
CH–6900 Massagno
Switzerland
Tel.: 41-91-571539. Fax: 4191 567865
and
Mr. Ove Poulsen
PKB OBp International Marketing
Hegnet 13
DK–2600 Glostrup
Denmark
Tel.: 45-42-965060. Fax: 45-42-965060

RESEARCH AND DEVELOPMENT

The Bf 110 V1, W.Nr. 868 powered by Jumo 210 engines. Initially that plane had trapezoid shape vertical tail later replaced by eliptical in E-stelle in Travemünde. It later got a designation D–AHOA. According to some sources it was flown first time in 12th of May, 1936 r. with dr Hermann Würster at the controls (not Rudolf Opitz as it stated in the other sources). *(MVT via M. Krzyżan)*

The creation of the Messerschmitt Bf 110 multi–role aircraft owed much to new... a fashion.

Of the mid–1930s, when work was started in France, Poland and Germany, on twin–engined multi–role aircraft, intended to replace light bombers as well as reconnaissance planes and fighters. The new type was to be also capable of escorting bombers into enemy territory, and one design was to fulfill all tasks previously assigned to multiple types.

In Poland the PZL P.38 *Wilk* (*Wolf*) was developed, while France designed the Potez 63. In Germany the RLM (Reichsluftfahrtministerium) issued in 1934 a specification for the so called Kampfzerstörer (bomber–destroyer). This was to perform the following types of missions:

– escorting bombers,
– intercepting and attacking enemy bombers,
– bombing attacks,
– attacking enemy ground forces,
– attacking enemy fighters.

The RLM expected a three–seat, twin–engined, all–metal low wing monoplane, armed with a 20 mm cannon and an internal bay for 500 kg of bombs.

This specification was circulated at the end of 1934 to the AGO (Arado), Dornier, Focke-Wulf, Heinkel, Henschel, Gothaer Waggonfabrik and BFW (Messerschmitt) aircraft companies. AGO (Arado) developed a heavily–armed (four 20 mm cannon) project designated Ao 225, while the Gothaer Waggonfabrik prepared the Projekt 3, a twin–boom aircraft similar to the Dutch Fokker G.I of 1937.

Eventually only Focke–Wulf, Henschel and BFW constructed prototypes. The first two prepared their projects strictly to the RLM specification, while BFW partly rejected the bombing role requirements.

Focke–Wulf built a large aircraft (25 m wing span!), powered by two DB 600 in–line engines rated at 700 kW (910 hp). The armament consisted of two fixed MG FF 20 mm cannons in the nose and another MG FF in the Mauser dorsal turret. The prototype Fw 57V1 was first flown in the Spring of 1936. It soon proved to be too heavy; without the armament the Fw 57 was heavier than the Henschel Hs 124 by as much as 164%! The prototype Fw 57V1 crashed in the summer of 1936 as a result of a DB 600 engine failure. At the same time the Fw 57V2 and V3 prototypes were under testing, which continued until the Autumn of 1936.

Another prototype was constructed by Henschel. The Hs 124 designed by Ing. Nicolaus was a three–seat, twin–engined mid–wing all–metal monoplane. It featured a Mauser nose turret with a single MG FF, manned by a gunner. The turret was electrically operated. The radio–operator was seated in the back, and he also manned

The Messerschmitt Bf 110 competitor – the Focke– Wulf 57V1. *(MVT via M. Krzyżan)*

Above: The Messerschmitt Bf 110 A–01 powered by Junkers Jumo 210 Da engine.
(P. Jarrett via B. Ketley)

Above: Flown in 1938 the Messerschmitt Bf 161V2 (CE+AT) would have been a prototype of reconnaissance and light bomber plane. It was ordered in 1935. The plane was never introduced into series production. Its crew would consist of two persons: pilot in single cockpit and observer/bombardier in glazed nose. Plane was powered by two Daimler Benz DB 601 engines. Clearly visible similar like in Bf 110 parts and design. *(P. Jarrett via B. Ketley)*

Below: The Messerschmitt Bf 110 B–2 fighters of II./ZG 1 *Wespen* w flying in formation, Germany 1939.
(P. Jarrett via B. Ketley)

an MG 15 machine gun. The prototype Hs 124V1 was powered by Junkers Jumo 210C in-line engines rated at 470 kW (640 hp). The second prototype, the Hs 124V2, was powered by the more powerful BMW 132Dc radial engines (licence–built Pratt & Whitney *Hornet*) rated at 640 kW (870 hp).

The competition was won by the third contender – the Bayerische Flugzeug Werke (BFW) – which developed the Messerschmitt Bf 110. The BFW team omitted some of the Kampfzerstörer requirements on purpose, as they considered that it was not possible to fulfill all of the planned mission roles (vide the Polish PZL P. 38 *Wilk*). It is worth noting here that the BFW was, at that time, fully occupied with developing the Messerschmitt Bf 109 fighter, and they could not commit themselves too extensively in developing a new sophisticated aircraft. For this reason Ing. Willy Messerschmitt and the design office manager, Ing. Walther Rethel developed their own specification. The BFW cancelled the internal bomb bay and reinforced the armament. This was supposed to substantially increase the performance, i. e. the range, maximum speed and climbing speed. At the same time, at the beginning of 1935, the RLM changed the original specification splitting it into two separate aircraft – the destroyer (Zerstörer) and fast bomber (Schnellbomber). The new requirements favoured speed, range and manoeuvrability, and these were met solely by the BFW design. BFW soon started work on the fast bomber under the designation of Messerschmitt Bf 162.

After the specification was changed, the Hs 124V3 was built, featuring a new, unglazed nose section and armed with two MG FF cannon and two MG 17 machine guns. This aircraft was used only for test purposes. The Messerschmitt Bf 110 was a slim, all–metal, cantilever low–wing monoplane of semi–monocoque design, with a retractable undercarriage and featured Handley Page automatic slats. To improve the rear–firing field the Bf 110 was fitted with twin vertical tail surfaces. The aircraft was powered by two DB

600A engines each rated at 700 kW (910 hp).

Upon completion of static and taxying trials at Augsburg–Haunstetten airfield, the prototype Bf 110V1 (W.Nr.868, registered D – AHOA) was first flown by Hermann Würster on May 12, 1936. The aircraft performed a circuit over the airfield, climbed to 2,000 m, and the control surfaces were thoroughly tested. The engines were found to cause some problems during the first flight, and the aircraft proved difficult to control at low to medium speeds. At high speeds it performed excellently. However, the manouevrability was inferior to that of the current biplane fighters (although superior to that of the Bf 109E–4!). Problems with ground manoeuvrability were also encountered while taxying which followed from the malfunctioning of the mainwheel brakes. On August 25, 1936 Würster (according to some sources, the pilot was Rudolph Opitz) flew the Bf 110V1 in front of Ernst Udet, the RLM technical chief, performing an aerobatic display. Later a mock dog – fight took place between a Bf 109 (V3?), flown by Udet himself versus the Bf 110. The 109 was not able to 'shoot down' the twin–engined Bf 110V1.

The second prototype Bf 110V2 (W.Nr. 869, D – AQYE) was first flown on October 24, 1936. The aircraft featured modified undercarriage and tail surfaces, and other minor refinements. On January 14, 1937 it was transferred to Erprobungsstelle Rechlin for tests. At Rechlin (between January 14 and 22) flight testing was carried out, including aerobatics, diving at angles of up to 20 degrees and take – offs and landings at maximum load. After the tests were completed, the BFW received an order for a pre–series batch of aircraft (so called Null Serie – zero–series). The Rechlin testing confirmed the BFW test pilots opinion that the aircraft needed some refinement. At the same time it was found that neither the Fw 57 nor the Hs 124 were suited for the destroyer tasks.

The Bf 110V3, third prototype, (W.Nr. 870, D – ATII) was first flown on 24 December 1936,

The experimental Messerschmitt Bf 110 C–04, assembled in Focke Wulf factory. It had an engine cooling liquid radiator transferred from underwing to underengine position and oil radiator from underengine to overengine position. Such solution was not applied in production planes – Messerschmitt left with previous design with engine cooling liquid radiators mounted underwing, similarly like in Bf 109.

(P. Jarrett via B. Ketley)

Above and below: The same plane in front and back view. On the above photo is clearly visible that there are not radiators under wings and wide underengine radiator inlet exist. On the below photo cooling air outlets from oil radiators mounted in overengine position are visible.

(Both MVT via M. Krzyżan)

The Messerschmitt Bf 110 C–4 – the most popular version of C variant, powered by DB 601A engines.
(MVT via M. Krzyżan)

this maiden flight lasting from 11:05 till 11:18 a.m. This was the first armed version, featuring four fixed Rheinmetall – Borsig MG 17 7.92 mm machine guns. The fourth prototype – Bf 110V4 (W.Nr. 910, D – AISY) was first flown on January 21, 1937.

The first production version, the Bf110A was to be powered by DB600A engines, but because of their malfunctions it was decided that a new version would go into production, powered by the Daimler Benz DB601 engines. The RLM did not expect the DB601 engines to be available before 1938, so the Bf 110A–0 airframes already in assembly were to be fitted with Junkers Jumo 210Da engines rated at 470 kW (640 hp), driving constant speed two–bladed metal airscrews. The first Bf 110A–01 was completed in August 1937. Compared to the Bf 110V1 – V3 prototypes, the Bf 110A featured longer, slimmer engine nacelles. The armament comprised four fixed MG 17 machine guns in the nose and a flexible MG 15 in the rear gunner's position. The Bf 110A–04 was completed in March 1938. The Bf 110A–0 aircraft reached a top speed of 430 km/h at an altitude of 3,800 m. Because of further problems with DB601 engines it was decided that a series of Bf 110B aircraft would be built, also powered by Junkers Jumo 210Da enignes. A new, aerodynamically refined nose section was developed which housed an improved armament of four MG 17 machine guns and two Rheinmetall – Borsig (Oerlikon) MG FF 20 mm cannon positioned at the bottom of the fuselage.

Two prototypes of the Bf 110B–0 version were built, the Bf 110V5 (W.Nr. 911, D–AAHI), first flown on March 15, 1938 and the Bf 110V7 (W.Nr. 917, D – AISY), first flown on April 19, 1938. This latter flight completed between 2:07 and 2:21 p. m. Both were flown by Hermann Würster. The aircraft were powered by Junkers Jumo 210Ga engines, rated at 500 kW (680 hp), fitted with two – speed superchargers and driving VDM–Hamilton constant–speed three-blade metal propellers. The pre–production machines were unarmed, the Bf 110B–1 series built from July on being the first version to have full armament fitted, including the MG 15 mounted on an Arado turntable. The Bf 110B was the first production version of the type.

The French Air Minister, Gen. Vuillemin's visit to the Messerschmitt AG works (the factory's name was changed during 1938) was the first occasion to display the new aircraft. During this visit to Augsburg, the French General was shown how the new Bf 110s left production lines in regular intervals, took–off and flew away. In fact the mistyfication lay in that after a short flight they landed nearby, were rolled inside the production hall, only to 'leave production line' once again, thus creating the impression of an advanced, large–scale production.

Besides the Bf 110B–1, also the Bf 110B–2 were manufactured, these having cameras but

Above: The Messerschmitt Bf 110 C–4 in factory painting.
(MVT via M. Krzyżan)

Below: The Messerschmitt Bf 110 C–4B. There is visible underfuselage mounted double bomb rack ETC 500, covered by special fairing.
(MVT via M. Krzyżan)

lacking cannons. A total of 45 Bf 110Bs were manufactured. During the war some of the Bf 110Bs were modified to the training Bf 110B–3 standard, with additional radio equipment for blind flying and navigational instruments.

In 1938 Bf 110Bs were delivered to the first destroyer units, I.(Z)/LG 1, I./ZG 1 (ex–JG 141) and I./ZG 76, where special Zestörerschulstaffeln were organized to facilitate pilot training. The Bf 110B–3s were used in flying training schools until 1941.

The Messerschmitt Bf 110Bs were also used as experimental machines. For example, the Rheinmetall–Borsig MK 101 30 mm cannon was tested in the Bf 110B–0 (D–AAPY), and Trommelgerät – 73 mm missiles, fired from tubular launchers, were tested on a Bf 110B (D–ADJD) at Tarnewitz on March 10, 1939. The missile launcher assembly was fitted in the fuselage nose, and consisted of eight individual launchers. The missiles were effective at distances of up to 300 m, with a hit probability of 40%. The tests were continued until 1943, but the RLM did not agree to introduce the missiles into service use.

Not until the end of 1938 did the Daimler Benz works at Stuttgart – Unterturkheim manufacture the first DB 601A engines rated at 810 kW (1,100 hp) which enabled start–up of the Bf 110C series production.

The Messerschmitt Bf 110C featured rear fuselage of increased stiffness, the forward radiators under the engines were replaced with flat underwing ones, and a fixed tailwheel was fitted, this already introduced in the later B series aircraft. Oil coolers were placed under the engine nacelles. To facilitate wing assembly, the wingtips were squared off, reducing the wing span by 600 mm. In February 1939 10 pre–series Messerschmitt Bf 110C–0s were delivered to the Luftwaffe, these soon followed by the production Bf 110C–1. The Bf 110C–1 reached a top speed of 538 km/h at 6,000 m, and standard range was 1,034 km. 159 Messerschmitt Bf

Above: The Messerschmitt Bf 110 E, designated CD+MO. *(P.Jarrett via B. Ketley)*

Above: The reconnaisance Messerschmitt Bf 110 C–5 (W.Nr 2177, 5F+CM) of 1./(F)14. That plane, previously damaged by British fighters, made a force landing in England, 21th of July, 1940. *(Denés Bernàd)*

Below: The Messerschmitt Bf 110 E–1 (BC+FG) with underfuselage mounted ETC 500/IXb bomb rack and and underwing mounted ETC 50 bomb racks. *(MVT via M. Krzyżan)*

The Bf 110 D–1/R1 on the assembly line. Note an underfuselage *Dackelbauch* (dachshund tank) additional tank.

110Cs were manufactured by September 1939.

It was a Messerschmitt Bf 110C that was involved in the thrilling attempt to steal an aircraft from the Augsburg factory. After Gen. Vuillemin's visit, the French intelligence service continued efforts to get a complete Bf 110 from the factory and fly it to France. The Chef des Services de Renseignement Air at Belfort, Col. Andre Serot, managed to recruit a former Luftwaffe pilot, Franz Xawer Öttil. Öttil was expelled from the service for causing an accident while serving in the StG 2 'Immelmann', when he crashed a Henschel Hs 123A while performing aerobatics at extremely low level. He found employment as a Start Monteur (Take–off fitter) at Messerschmitt AG at Augsburg. On May 10, 1939, during the lunch break, he hi–jacked a factory–fresh Messerschmitt Bf 110C–1 (W.Nr. 979). After a short flight he landed in a field at Bad Wörishofen, where his older brother waited for him with large quantity of fuel in canisters. After refuelling they took off for France. The aircraft crashed at Villers-sous-Chalamont, some 100 km from Belfort, because of bad weather. While looking for a place to land Öttil flew the plane too low, and hit a hill. Both brothers were killed in the crash. The engines and other elements were later investigated by French experts from *Inspection Generale Technique de l'Air*.

Messerschmitt Bf 110Cs were also manufactured by the Focke–Wulf at Bremen, Gothaer Waggonfabrik (GWF) at Gotha and MIAG (Mühlenbau Industrie AG) at Braunschweig. They were also used for testing.

The Messerschmitt Bf 110C–2 differed from the earlier variant in that it had the FuG 10 radio set instead of the FuG IIIa, and a modified gunner's position. The Messerschmitt Bf 110C–3 was armed with Rheinmetall–Borsig (Oerlikon) MG FF/M 20 mm cannon, featuring a rear loading system. The Bf 110C–4 was the first version with seat armour for the pilot and gunner.

Another variant of this production version existed, this being the Bf 110C–4B Jabo (Jagdbomber – fighter–bomber), powered with Daimler Benz DB 601N engines rated at 920 kW (1250 hp). These were fed with 96 – octane C3 leaded fuel, rather than the standard 87 – octane B3 fuel. The Bf 110C–4B featured two ETC 500/IXb bomb racks under the fuselage for 250 kg or 500 kg bombs. The Messerschmitt Bf 110C–4B Jabo were used by the E.Gr 210 experimental unit.

The Messerschmitt Bf 110C–5 was a reconnaissance version, fitted with the Robot Rb 50/30 camera for vertical photography, positioned inside the fuselage, behind the pilot's seat. The lens was placed centrally under the fuselage and protected with a moving cover. The Bf 110C–5 armament was reduced to four Rheinmetall – Borsig MG 17 7.92 mm machine guns in the nose and a single MG 15 in the rear gunner station. The Messerschmitt Bf 110C–5N was powered with DB 601N engines.

The final versions of the Messerschmitt Bf 110C were: the Bf 110C–6, armed with a single Rheinmetall – Borsig MK 101 30 mm cannon in a streamlined pod in the nose, rather than the two MG FF or MG FF/M cannon; and the Bf 110C–7, powered with DB 601N engines, fitted with two ETC 500/IXb racks. The Bf 110C–7 were modified with extended main undercarriage legs to enable the carriage of 500 kg bombs. The Bf 110C–4B Jabo was able to carry two 500 kg bombs on the ETC 500/IXb rack, but because of the small ground clearance this configuration was not used. KE+UK, first flown at Rechlin on July 11, 1940 was the Bf 110C–7 prototype.

Many other Bf 110 aircraft were used for various tests. The Bf 110B–0 (D – AAPY) was used in July/August 1939 for testing of the MK 101 cannon. The Bf 110C–1 (W.Nr. 974, BA+CP) was fitted experimentally with two FDL 81Z remotely–controlled firing stations (Ferngerichte Drehringlafette), positioned under the fuselage and at the rear of the cockpit. Each station featured twinned Mauser MG 81Z 7.92 mm guns.

The Bf 110C–1, W.Nr. 945, GK+AY, the personal mount of Gen. Ernst Udet, was an interesting machine. The aircraft had the rearview mirror fitted inside, rather than on the outside of the cockpit. The explanation is that Udet liked to have a drink of brandy during flight, and he wanted to know if his engineer, Gefr. Kurt Schnittke did the same. The next Udet mount, the Bf 110D–0, W.Nr. 3354, VF+HP, was similarly equipped.

During the Norwegian campaign the range of Messerschmitt Bf 110C was found to be insufficient. Additional fuel tanks were necessary. The first modification consisted of adding, under the fuselage a large plywood tank for 1,200 ltr. of fuel, this called the Dackelbauch (dachshund–tank). The langstrecken, or long range, aircraft were initially designated Bf 110D–0. After the series production was started, the designation was changed to Bf 110D–1/R1.

The Bf 110D–1/R1 entry into service was not a successful one. The tanks suffered from air drop mechanism failures, and a direct hit resulted in explosion of the aeroplane (e. g. that of Hptm. K. Restemeyer, the C. O. of I./ZG 76). Several explosions happened also because of poor tank ventilation leading to fuel vapour being collected inside. If we mention that the tank also suffered from insufficient stiffness, it is easy to understand why another version was soon developed, the Bf 110D–1/R2, equipped with two 900 litre metal streamlined tanks, fitted under the outer wing panels. The tanks, developed by Junkers, could be jettissoned in flight. The Bf 110D–2 featured hatches for two 300 litre fuel tanks

The Messerschmitt Bf 110V19 (modernised production G–4) with kit of 12 *Bordrakete* RZ 65 (RZ – *Rauch Zylinder*). During tests were proved unusability of this kind of armament.

(MVT via M. Krzyżan)

The Bf 110 C–1 (BA+CP) FA 12 (*Fernantrieb* 12) with experimental remote controlled *Ferngerichte Drehringlafette* FDL 81Z turrets with doubled MG 81Z machine guns.

(MVT via M. Krzyżan)

and two ETC 500/IXb bomb racks under the fuselage (the long range fighter – bomber version). The Bf 110D–3 had a larger rear fuselage section which housed a dinghy and an additional 75 litre underfuselage oil tank. It was capable of carrying additional 300 or 900 litre fuel tanks; the ETC 500/IXb underfuselage racks could also be fitted.

During 1939 and early 1940, Allied bombers seldom appeared over Germany. The few day bombers were easily intercepted and destroyed by the Bf 109 and Bf 110 fighters. The situation changed dramatically when the night raids commenced; the few German fighters used in the night–fighting role (the Arado Ar 68E and Bf 109C and D) were not well suited for actions in the darkness, since they had no homing (radar) devices and their range was limited. It was necessary to introduce multiengined aircraft, with a longer endurance. Initially the standard Bf 110 Cs were allocated to Nachtjagd (night fighter) units, while in the Autumn 1940 the Bf 110D–1/U1 appeared with the infrared radiation detectors (so called Spanner–Anlage). The device enabled positioning of an aircraft with use of infrared radiation source detection (like that of exhaust pipes). The infrared detector was placed under the forward fuselage. Upon detection of a radiation source (an aeroplane) a small bright band appeared on the little round screen (so called Q–Rohr) in front of the cockpit. The night fighters also featured exhaust flame and flame screens to protect the crew from the glare.

At the end of 1930s the Augsburg works of the Messerschmitt AG started work on the Me 210, supposed to supersede the Bf 110. Unfortunately, the aircraft design met serious troubles (like fuselage cracking) and thus the new type could not enter Luftwaffe service in 1941 as planned. Because of the planned reequipment, the Bf 110 production was reduced from 1,231 aircraft in 1940 to 786 the following year. The production of Bf 110 at Focke–Wulf GmbH at Bremen was terminated in July 1941. Thus the GWF GmbH at Gotha and MIAG at Braunschweig, remained the principal Bf 110 manufacturers, and during 1942 these assembled a total of 468 aircraft.

In the Spring of 1941 Messerschmitt developed a new version, designated the Bf 110E. The Bf 110E–0 pre–production series aircraft and the early Bf 110E–1s were powered by the DB 601A–1 engines, the remaining machines using the DB 601Ns. The Messerschmitt Bf 110E–1 was manufactured in two variants: the E–1/U1, a night fighter with the Spanner – Anlage device; and the E–1/U2, with a redesigned, three–seat cockpit, with a place for the Leitoffizier (navigator/homing device operator) behind the pilot.

The Bf 110E differed from the earlier series in that they featured four ETC 50/VIIId bomb racks for 50 kg bombs (e.g. SC 50 or SD 50), placed in pairs under outer wing panels. The Bf 110E–2 had similar modifications to those of the Bf 110D–3. The Bf 110E–3 was a long range reconnaissance version lacking the MG FF cannon or the underfuselage ETC 500/IXb bomb rack; it could take the additional underwing fuel tanks (300– or 900–litre types). Some Bf 110E–3s had two rear–firing MG 17 machine guns mounted in the rear fuselage. The photographic equipment comprised one Rb 50/30 camera.

The Bf 110F version was powered with Daimler Benz DB 601F engines with take off rating of 976 kW (1,328 hp). The pre – production Bf 110F–0 was identical (apart from the engines) to the Bf 110E–0. Production Bf 110F–1s had enlarged oil coolers (the Fö 812 type) under the engines. An armoured windscreen plate 57 mm thick was standard, as was 8 mm armouring to the pilot and gunner positions. The pilot's seat was protected with 12 mm armour plate; the side panels of the cockpit glazing were 35 mm thick, while the top panel of the gunner's cockpit was 57 mm thick. The Bf 110F–1 had two ETC 500/IXb racks under the fuselage and four ETC 50/VIIId under the wings. The bomb racks were removed in the Bf 110F–2. At the Waffenprüfplatz (weapon test range) at Tarnewitz, two Wr.Gr 21 210 mm tubular missile launchers under the wings were tested. Bf 110V19 (W. Nr 2656), the prototype for the Bf 110F–2, was also used for tests of the RZ65 unguided air–to–ground missiles; the missiles were placed in an underfuselage launcher (12 connected 65 mm tubular launchers). The tests at Tarnewitz were carried out between October 5, 1942 and April 14, 1943. The Bf 110F–3 was equivalent to the Bf 110E–3.

The Bf 110F–4 was a night fighter version; these aircraft of featured enlarged vertical tails and positioned trim tabs. The modifications resulted from test flights with one engine off. Armament was improved by adding two Rheinmetall–Borsig MK 108 30 mm cannon placed in an under–fuselage pack (100 rounds per cannon). In 1943 the cannons were removed; in some aircraft (such as the Bf 110F–4/U1) these were installed again as the 'Schräge Musik' assembly, in the cockpit, to fire obliquely upwards. The pilot used a modified Revi C/12D gunsight. In some aircraft armed with 'Schräge Musik', the rear MG 15 was removed. The Messerschmitt Bf 110F–4 could be equipped with *Lichtenstein* BC (FuG 202) radar, with the designation Bf 110F–4a. Also, the armament was changed: the MG FF/M cannon were replaced with Mauser MG

The cockpit of Bf 110D with visible *Revi* reflection gunsight and ZFR tube gunsight.

(MVT via M. Krzyżan)

The Bf 110 E–1/U1 of 6./ZG 1 (2J+AP) with three peoples of crew. This plane serviced in 1942 in Eastern Front.

(P. Jarrett via B. Ketley)

151/20 20 mm cannon (300 rounds for the port cannon and 360 rounds for the starboard one). The BC radar (FuG 202) was using a four–dipole antennae, the so called Maikäferfhler. Two additional 300–litre fuel tanks were fitted under the wings. The Bf 110F–4 (and F–4a) were manufactured at the MIAG and GWF factories.

The Messerschmit Bf 110G owed its existence to the successful marriage of the new Daimler Benz DB 605 engine to the Bf 109, the standard Luftwaffe day fighter; and to the serious problems with bringing the Me 210, the Bf 110's successor, into production. In June 1941 W. Messerschmitt offered to modify the Bf 110F by installing the Daimler Benz DB 605B–1 engines, rated at 1,085 kW (1,475 hp). Initially the RLM was not too enthusiastic about the idea, but in January 1942, when the Me 210 series production startup was still being delayed, the RLM decided to start Bf 110G manufacturing. Several versions were planned: the G–1 Jabo (fighter–bomber); the G–2 Zerstörer (destroyer, or heavy fighter); and the G–3 Fernaufklärer (long range reconnaissance plane). The Bf 110G featured modified engine cowlings and propeller spinners, similar to those on Me 210 and Bf 109G; identical spinners were also used on some late Bf 110F–4s.

Because of overloading with work on Bf 109G production preparation, the first Bf 110G–0 did not leave the factory before May 1942; series production was started in December 1942. The Bf 110G–2 was the first version (the Bf 110G–1 was cancelled). The first Bf 110G–2s were powered with DB 605B–1 or DB 605 A–1 engines, rated at 1,085 kW (1,475 hp) driving VDM9–12078A/B propellers. The armament was similar to that of the Bf 110F–2, i. e. 4 x MG 17, 2 x MG FF/M and an MG 15. Starting from February 1943 changes were introduced: the FuG 10P radio set (with the Peil G.VI) which replaced the previous FuG 10, and the FuG 25a IFF was introduced in place of the FuG 25. Also the armament was modified: the MG 15 was replaced with Mauser MG 81Z 7.92 mm twin machine guns (750 rounds per gun), and the MG FF/M with the MG 151/20 cannon (400 rounds for the port, and 350 rounds for the starboard gun). In the Summer of 1943 in aircraft used in the Reich air defence system the FuG 10 radio set was replaced by the FuG 16.

The Messerschmitt Bf 110G–2 was also fitted with an additional 75 litre oil tank and a 3–litre tank for ether–petrol, which was used for running up a cold engine. The Bf 110G–3s featured the Rb 50/30 camera and FuG 17 radio set, tuned for communication with ground units. The Bf 110G–4, like the Bf 110F–4, was planned as a night fighter. Its production started in January 1943 at GWF and at Luther Werke GmbH at Braunschweig. Up to Fabruary 22, 1945 a total of 1,850 aircraft of this version were manufactured.

During the first montns of Bf 110G–4 use many malfunctions of the DB 605 engines were encountered, often causing fires. The engines were tested thoroughly at the Erprobungsstelle (test centre) at Rechlin. It was found that the fires started because of engine seizure when flying at high altitude; the Messerschmitt AG specialists found that the malfunction was due to aeration of oil caused by faulty working of the engine supercharger. Daimler Benz, the engine manufacturer, suggested an increase in oil tank capacity. Subsequent modifications were introduced to the DB 605 engine, including the exhaust piping and the fuel system. The equipment was continuously improved, e. g. the FuG 16ZE and FuG 10P radio sets were intrduced. Also the armament was strengthened: the aircraft carried two MK 108 cannon in the nose, two MG 151/20 in a container under the fuselage, two upwards firing MG FF in the cockpit and the gunner's rear–firing MG 81Z.

When fighting the B–17 *Flying Fortress* and the B–24 *Liberator*, their large caliber machine guns proved deadly weapons and armour reinforcement was found necessary on the Bf 110Gs. Cockpit sides were armoured with 5 mm panels, and the cockpit glazing was made entirely of 35 mm bulletproof glass panels. The night–fighting models featured Eberspächer (flame dampers) on the exhaust stubs. The later production aircraft were fitted with modified radar and radio equipment (the FuG 16Zy, FuBl 2RL5 and FuG 101). The FuG 220 radar installation necessitated increase to the fin (by 0.5 sq m) and rudder (by 0.7 sq m) surfaces.

The Bf 110G–4a, G–4b etc. designations were applied to distinguish various types of radar equipment fitted to the Bf 110s.

The most modern radar fitted to the Bf 110 was the *Lichtenstein* BC (FuG 202), developed by Telefunken. This worked in the 490 MHz frequency with a pulse power of 1.5 kW. It featured a very sensitive super–reactive receiver with a six–step amplifier. The antenna had four dipoles and reflectors, fitted to the nose part of the fuselage to form a square. These could either work in the receiving and transmitting channel, switched with a capacitive controlled motor. Prior to transmission the pulse was modified by a phase convertor to produce a narrow conical transmisson characteristic. In the receiving channel, the antenna dipoles were so connected, through a *Quirl* device, to the monitoring screens, that they acted on the port and starboard bearing indicator deflecting electrode and the upper and lower altitude indicator electrode. The monitoring set consisted of three screens (indicators), i.e. the bearing and altitude indicators mentioned above, the former having a horizontal time base and the latter a vertical one, these two indicators showing the selected target at a distance of up to 1.5 km; and the third screen which showed all the targets within the range of 8 km an which served to pick the interception target. The radar was successfully tested during August and September 1941 on the Oblt. Becker's Do 215 B–5 board, who claimed 6 Allied bombers before he was shot down himself. After introducing some refinements, the RLM ordered series production of the radar by the Telefunken in February 1942. Later modifications to the system did not introduce any improvement, and in fact they only led to worsening its performance. In spite of that Dr Muth and Dr Urtel presented a new *Lichtenstein* C1 (FuG 212) radar project. This was similar to the FuG 202, worked in the same 490 MHz frequency and had the same antenna array. The transmitter, recei-

The same plane in little different view. Two ETC 500/IXb bomb racks with common fairing under fuselage and underwing ETC 50 bomb racks visible.

(P. Jarrett via B. Ketley)

ver, and screen set were, however completely new. During May 1942 it was tested with result far from revelation. Despite the fact that the refinements seemed to lead in a direction opposite to that intended, the FuG 212 was put into production. It was used by a limited number of night fighting Bf 110G–4s.

Later, a combination of the good resolution of the *Lichtenstein* C1 and the wide–angle performance of the *Lichtenstein* BC led to the creation during the summer of 1943 of a new radar, called *Lichtenstein* SN2 (FuG 220). This had the transmitter and receiver of the earlier *Lichtenstein* S (FuG 213) intended for surface vessel detection; and worked in the 91 MHz frequency. There were two screens, the first giving the target bearing, the other giving the relative altitude. Several variants of the radar were soon developed, which differed only in antennae and operation modes. The most often used was the 'b' antennae arrangement fitted on the nose to form a square of four vertical dipoles. This was similar to that of the FuG 202, but much larger. Also, the electronics of the radar were subject to modification, to make it invulnerable to countermeasures. The 'd' version was to have a reversed antenna layout to search the zone behind the aircraft, but this was soon abandoned. Later on, the 'd' designation was given to the layout where the dipoles were tilted at an angle of some 45 degrees. The FuG 220 SN 2 radar would most often cooperate with the FuG 212 C1, so most night fighting Bf 110G–4s had two antenna systems on the nose – the smaller FuG 220 SN2, and the larger for the FuG 212 C1.

In 1943, after inspecting the British *Monica* radar taken from a bomber that crashed at Rosendahl in Holland, a radar called *Rosendahl-Halbe* (FuG 221a) was created. This worked in the frequency range of 190 to 220 MHz, and detected *Monica* transmissions. It had an antennae system consisting of four horizontal dipoles mounted on two extensions on the nose of the Bf 110G. It was used for a short time, being phased out by more modern warning radars. Later, the *Flensburg* (FuG 227) system was mounted on Bf 110Gs, this was a passive warning radar that detected the *Monica* transmission and the *Naxos* Z (FuG 350 Z) system which detected the H_2S beam. These passive radars were used to home on bombers that used the respective active radars. They gave long detection ranges of up to 50 – 72 km, thanks to detection of even weak beams well beyond the *Monica* and H_2S detection range. The specification of these radars is shown in the table.

On December 15, 1943, during a meeting at GWF Gotha, a specification was laid out for a new version called the Messerschmitt Bf 110H. This was to be powered with Daimler Benz DB 605E engines rated at 1,270 kW (1,730 hp), and was to incorporate many modifications:

– a redesigned cockpit cover with the hood jettisoned in emergency;

– a redesigned tail with a retractable tailwheel;

– a longer fuselage, enlarged tail surfaces (made of wood), enlarged wing with rounded tips (like the Bf 109F to K);

– a redesigned main undercarriage;

– DB 605E engines with Fö 870 oil coolers and enlarged radiators;

– modified armament.

Several variants were planned: the Bf 110H–1 Jabo, Bf 110H–2 Zerstörer (armed with two MK 108s and one MK 103 30 mm cannons plus an MG 81Z), Bf 110 H–4 Nachtjagd (avionics including: the FuG 220, FuG 16ZY, FuG 135, FuG 101) and the Bf 110 H–5 (single–seat heavy fighter). In the Autumn of 1943 construction was started at Luther Werke of the wooden wing tips and the longer wing itself. The prototype was to be first flown during March 1944. On February 24 the GWF suffered an air raid by the USAAF 8th Air Force, which led to destruction of the design office and the prototype workshops. The Bf 110G fitted with DB 605BS engines (DB 605 with DB 603–type supercharger), planned to test the Bf 110H modifications, was destroyed. In 1944 (March 11 or April 15) the RLM ordered continuation of work on Bf 110H; the scope of modifications was limited to wing redesign and fitting of the DB 605E engines. During September, work on the cockpit was discontinued. After pronouncing the Jagernotprogramm (emergency fighter programme) the Bf 110G production was reduced and the Bf 110H was abandoned.

The Bf 110C to G could accomodate the tropical equipment (Tropenausrüstung) consisting of air filters on the carburettor air intakes (the air intake for the port engine was positioned on the port wing leading edge, while that for the starboard engine on the engine cowling), additional food and water rations, and a gun for animal protection (this was placed between the 9th and 11th fuselage section), an enlarged oil cooler (in the Bf 110F–2 Trop this was the SKF Fö 802) and an additional armament cover.

The Bf 110 was subject to numerous modifications, so called Rüststze and Umrüst–Bausätze. The Umrüst – Bausätze were fitted in production or during general overhaul.

These were:

– Bf 110C–1/U1: a towing attachment for the Messerschmitt Me 321 Gigant glider; the towing team consisted of three Bf 110C–1/U1s, the so called Troika–Schlepp;

– Bf 110D to E–1/U1: Spanner–Anlage, in some late Bf 110E–1/U1s manufactured with the FuG 202;

– Bf 110G–2/U1 and G–3/U1: changed armament; G–2 – MG FF/M to MG 151/20 or Robot Rb 50/30 camera; G–3 – to MG 151/20:

– Bf 110E–1/U2: a place for the third crew member, the Leitoffizier;

– Bf 110C to E/U3: additional radiator armouring;

– Bf 110 to E/U4: additional armouring of the radiators, oil coolers and the bottom of engine nacelles;

– Bf 110G/U5: reconstruction of the forward fuselage to accomodate the FuG 212 C–1 radar antenna with dipole tilt angle control within the range of 70 and 120 degrees, which enabled operating mode to be changed from maximum of 4,000 m to 2,000 m and from 200 m to 100 m;

– Bf 110C to E/U6: armouring of forward engine naceles.

– Bf 110G/U6: probably the combination of U5

The captured Messerschmitt Bf 110 G–4c/R3 (B4+KA) with FuG 220 *Lichtenstein* SN2 radar. *(P.Jarrett via B. Ketley)*

antenna with FuG 221a *Rosendahl–Halbe* radar; the device was used to detect the British *Monica* tail warning radar.

– Bf 110G/U7 (later changed to R2) – GM1 installation (explained unofficially as Göring Mischung – Göring Mixture) for liquid N_2O nitrous–oxide (so called laughing gas) injection into the engine supercharger which enabled the engine power to be maintained when flying at high altitude. The system ensured sufficient oxygen to boost the engine by over 200 kW (some 280 hp) for 19 minutes. The cylindric nitrous–oxide reservoir (1,600 m long, 650 mm in diameter, weighing 307 kg) was fitted in the rear gunner cockpit. So equipped was for example the Bf 110 G–4/U7 W.Nr. 5457, NM+SE. A few U7 aircraft reached night fighting units at the end of 1943, e. g. D5+AD (W.Nr. 5560) of III./NJG 3;

– Bf 110G/U8: an additional 541 litre fuel tank in the gunner's cockpit and a 35 litre oil tank , which enabled the range to increase to 1,270 km, or even 1,620 km (with two 300–litre drop tanks). The use of the tank negatively affected the Bf 110G flying performance.

– Bf 110G/U9 (later designated R3): two Rheinmetall MK 108 cannon in place of the MG 17s (120 rounds for the port cannon and 135 rounds for the starboard one). The modification was

The Messerschmitt Bf 110 G–4b/R3 with FuG 212 i FuG 220b radar antennaes. Under fuselage visible 'tank' with additional armament – two MG 151/20 guns. *(MVT via M. Krzyżan)*

carried out at the Posener Werk Abteilung at Poznań–Ławica in late 1942 (December 9 ?) using the 0504 and 0505 cannons. Test flights took place on March 18 and 19, 1943. From May 21, 1943 the GWF started production of the Bf 110G–4/U9. They entered service on August 4, 1943. Serious problems with the cartridge case chute were encountered,and the barrel flame of MK 108 practically melted one of the FuG 202 antennae. When using the M1 modification it was necessary to protect the underfuselage MG 151/20 pack from the MK 108 cannon cartridge cases,and it was advised to use a barrel flame damper. In December 1943 the designation was changed to R3.

The MK 108 cannon were not favoured by the crews; they claimed that at night the attack was at short range, and the MK 108 was not necessary.

The top night fighter pilots, such as Jabs, Schnaufer, preferred the MG 151/20 installation.

Rüstsätze were modifications carried out in the field from complete kits delivered for fitment. These were:
– Bf 110D/R1: additional drop fuel tanks; under the fuselage for 1,010 litres, or under the wings for 900 or 300 litres;
– Bf 110E/R1: additional armouring for the pilot and gunner's cockpit sides;
– Bf 110G/R1: a BK 3.7 cm Flak 18 cannon mounted in a ventral pack. The ammunition magazine was fitted in place of the removed MG 151/20 ammunition drums. The ammunition reserve was 66 rounds in 11 magazines. The Bf 110G/R1 were tested in the Versuchskommando für Panzerkampfung at Rechlin. The Bf 110G/R1 were used among others by 5./ZG 1. The Erprobungskommando 25 used the Bf 110G/R1s in attacks against B–17s and B–24s.
– Bf 110D/R2: 900– or 300–litre fuel drop tanks and additional 75 litre oil tank under the fuselage.
– Bf 110E/R2 – identical to the Bf 110E/R1.
– Bf 110G/R2 = Bf 110G/U7.
– Bf 110E/R3 – additional armouring.
– Bf 110G/R3 = Bf 110G/U9.
– Bf 110E/R4 – see Bf 110E/R3.
– Bf 110G/R4 = R2 + R3.

Willingly used the Bf 110 G–4/R3 modification, where MK 108 cannons were replaced by MG 151/20 cannons with fire dampers.

– Bf 110G/R5 = R1 + R3.
– Bf 110G/R6 = R2 + R3, the rear gunner's position with the MG 81Z retained, GM1 tank moved forward, the gunner in a crouched position. The prototype Bf 110G–4, (W.Nr. 730059), coded TJ+ZO (?) was first flown on May 13, 1944.
– Bf 110G/R7: standard fitting of two additional 300–litre tanks.
– Bf 110G/R8 = R3 + 'Schräge Musik' (2 x MG/FF/M); the prototype developed by Ofw. Paul Mahle of II./NJG 5. The gunsight was on the cockpit cover, and the aerial mast repositioned to the side.
– Bf 110G/R9: two Rheinmetall–Borsig MK 108 cannons mounted in the cockpit as the 'Schräge Musik'. Not used in production models, planned for the Bf 110H.

Other modifications:
– Bf 110F/B1 and Bf 110G/B1: additional 75 litre oil tank.
– Bf 110G/B2: additional fuel tanks.

M type modifications:
– Bf 110G/M1: two underfuselage Mauser MG 151/20 cannon packs, with 200 rounds per cannon.
– Bf 110G/M2: two ETC 500/IXb bomb racks under the fuselage (most often the SC, SD 250 or SC, SD 500 bombs).
– Bf 110G/M3: four ETC 50/VIIId bomb racks under the wings (SC, SD 50, SD 70 or SD 50F bombs, SdC 10 pod or LC 50F flares).
– Bf 110G/M5: two W.Gr 42 210 mm unguided rocket launchers.
– Bf 110G/M4: two SD 2/XII launchers under the outer wing panels, each container with 24 SD2 splinter bombs.
– Bf 110G/M?: two MG 151/20 cannons instead of the MG 17.
– Bf 110G(?)/M?: two M1–type packs (modified) with MG 151/20 under the outer wing panels. A photo is known of an aircraft of ZG 1 with such armament, at the Wels airfield in May 1944.

The Messerschmitt Bf 110 G–4B/R3 nose with antennaes of FuG 212 *Lichtenstein* C1 (small dipoles) and FuG 220 *Lichtenstein* SN2b (big dipoles) radars.
(MVT via M. Krzyżan)

MESSERSCHMITT Bf 110 IN SERVICE

The Messerschmitt Bf 110 C–4/N in flight. The plane is in factory markings.
(P. Jarrett via B. Ketley)

BATTLE OF POLAND – 1939.

The Messerschmitt Bf 110s were first used operationally during the Polish campaign, the so called 'Fall Weiss'. On August 31, 1939 Luftwaffe combat and training units possessed 27 Bf 110Bs, of which 23 were serviceable, and 68 Bf 110Cs, of these 59 serviceable. Together this formed about half of the total number of the Bf 110s manufactured up to that time.Three Gruppen (squadrons) of Bf 110C–2s were used against Polish aviation. Two of these were within Luftflotte 1, and the third in Luftflotte 4. The latter was I./ZG 76 under Hptm. Werner Reinecke. The unit was based at Ohlau (now Oława, Poland) and comprised 35 aircraft of which, as of 9:00 a.m. on September 2, 1939, 31 were serviceable. Luftflotte 1 had I./ZG 1 under Maj. Hüth at Mühlen (now Mielno, Poland) with 34 aircraft (27 serviceable) and I.(Z)/LG 1 at Jessau near Königsberg (now Kaliningrad, Russia) under Maj. Walter Grabmann. The Gruppe had 33 planes, 32 of which were serviceable.The Messerschmitt Bf 110 – equipped units were given the task of escorting bombers, and flying offensive sweeps to draw into combat and destroy Polish fighters, as well as to attack Polish bombers and reconnaissance planes.The first to enter combat were the Bf 110C–1s of 2./ZG 76, led by Olt. Wolfgang Falck, which escorted the Heinkel He 111Ps of I. and III./KG 4 'General Wever' in their attack against Cracow's Rakowice airfield. During the sortie Falck attacked by mistake a Junkers Ju 87B–1 of the I./StG 2 'Immelmann' which 'got mixed' with the Messerschmitts; the attack was caused by confusing a red code letter with the red–and–white Polish national insignia.

The Bf 110Cs of the Luftflotte 1 were also very active; their first task was to escort the Heinkel He 111Ps of KG 27 'Boelcke' to Warsaw. During that mission they were intercepted by Polish fighters of the Brygada Pościgowa (Pursuit Brigade). A Messerschmitt Bf 110C of I.(Z)/LG 1, flown by Gefr. Warrelmann shot down the PZL P.11c of ppor. (P/O) Jerzy Palusiński of 111 Eskadra Myśliwska (111 Fighter Flight) of Brygada Pościgowa. During that fight the I.(Z)/LG 1 C.O., Maj. Grabmann was wounded. During the afternoon mission over Warsaw the Bf 110s made a surprise attack against PZL P.7a fighters of the 123 Eskadra Myśliwska, and The Bf 110s shot down four

The Messerschmitt Bf 110 C–2, 1939/40.
(MVT via M. Krzyżan)

The Messerschmitt Bf 110 C–4 in flight. *(P. Jarrett via B. Ketley)*

Polish aircraft of the 123 EM and the P.11 of por. (F/O) Aleksander Gabszewicz of 114 Eskadra. The kills were claimed by Gefr. Warrelmann, Hptm. Werner Schleif (Grabmann's deputy) and Ofw. Herbert Schob.

On September 2 the Messerschmitt Bf 110Cs of 2./ZG 76 met over Łódź the PZL P.11 fighters of the 161 Eskadra Myśliwska. Lt. Helmut Lent and Olt. Hans Nagel shot down two Polish fighters, flown by ppor. Jan Dzwonek and pchor. (C/O) Edward Kramarski. The Polish pilots proved tough opponents, shooting down two Messerschmitts. One fell prey to ppor. Tadeusz Koc of 161 Eskadra and ppor. Czesław Główczyński of 162 Eskadra, while the other was shot down by kpr. (LAC) Jan Malinowski of 162 Eskadra. The combat took place over Pabianice.

On September 3, the I.(Z)/LG 1 lost a single Bf 110, shot down by Polish fighters. The next day por. (F/O) Arsen Cebrzyński of the command section of III/1 Dywizjon Myśliwski (III/1 Fighter Squadron) of Brygada Pościgowa shot down a Bf 110. This was probably a I./ZG 1 machine, as the I.(Z)/LG 1 flew no sorties on that day.

On September 3, 1939, plut. (Cpl.) Marian Bełc of 152 Eskadra of the 'Modlin' Army aviation, shot down a Bf 110C of I.(Z)/LG 1 at Ciechanów. Two Messerschmitts were claimed by the pilots of 132 Eskadra Myśliwska of the 'Poznań' Army aviation: on September 7 at 7:00 a. m. in the Kalisz–Koło area pchor. Jan Maliński shot down a Bf 110–(but was himself shot down), and the next day kpt. (F/L) Franciszek Jastrzębski also shot down a Bf 110C. Both Messerschmitts were from the I./ZG1. On September 4, 1939 kpt. Mirosław Leśniewski, the C.O. of 142 Eskadra Myśliwska of the 'Pomorze' Army aviation shot down a Bf 110C over the airfield at Poczałków. During the combat kpt. Leśniewski's aircraft was hit and he was forced to land at Bramowo. Several other Bf 110Cs were destroyed by the Brygada Pościgowa. On September 3, 1939 at about 10:00 a. m. Polish pilots shot down two Bf 110s. In the afternoon the Messerschmitts took their revenge; over Okuniew the P.11 of ppor. Mirosław Ferić was shot down while the P.11s of pchor. Janusz Maciński and st.szer. (AC1) Henryk Szope were forced to land after being hit. Both Polish fighters were credited to Gefr. Warrelmann. On September 6 three pilots of III/1 Dywizjon, kpt. Zdzisław Krasnodębski, por. Cebrzyński and ppor. Ferić claimed a Bf 110 of I.(Z)/LG 1, while ppor. Hieronim Dudwał of 113 Eskadra Myśliwska shot down a Bf 110 of I.(Z)/LG 1.

The Messerschmitt Bf 110 C–4/N on the front airfield. *(MVT via M. Krzyżan)*

The Bf 110 C–4 of 8./ZG 76, France, 1940.
(MVT via M. Krzyżan)

Near Zaborów a group of six Polish fighters shot down another Bf 110C, this time of I./ZG 76. The same day ppor. Wiktor Strzembosz of 112 Eskadra shot down a Bf 110C. The crew baled out and hid in the forest. After six days the airmen returned to their unit. A Messerschmitt Bf 110C, flown by Ofw. Herbert Schob shot down a P.11 of st.szer. Leon Nowak of 112 Eskadra Myśliwska.

On September 5, 1939 Hptm. Wolfgang Falck of I./ZG76 shot down a Fokker FVII/3m transport of 211 (11) Eskadra Bombowa (Bomber Flight) of Brygada Bombowa (Bomber Brigade), while the next day the Messerschmitts destroyed a PZL P.37B *Łoś* bomber of the 212 (12) Eskadra bombowa at Kuciny airfield. The I./ZG76 aircraft also attacked Polish aircraft at Dęblin air base, where they destroyed about a dozen training aircraft (the combat report mentions 11 aircraft). On September 5, Falck shot down a PZL P.23 *Karaś* of 31 Eskadra Rozpoznawcza (Recconnaissance Flight) of the *Łódź* Army aviation, while Lt. Gustav Üllenbeck of I./ZG76 – a *Karaś* of the Brygada Bombowa (Bomber Brigade).

During the Polish campaign the Luftwaffe lost a total of 12 Messerschmitt Bf 110Cs. Gefr. Warrelmann of I.(Z)/LG1 was the top scoring Bf 110 pilot in Poland. This Gruppe claimed no less than 30 Polish aircraft. In fact, all Bf 110 units between them shot down no more than 15–20 aircraft over Poland.

WESTERN FRONT

After the war broke out, all the combat units equipped with Messerschmitt Bf 110Cs were sent to Poland. The Western Front saw only the destroyer (Zerstörer) units in formation, temporarily equipped with single–engined Bf 109s, mainly the Bf 109Ds. In the Autumn of 1939 the I./ZG 26 was re–equipped with Bf 110Cs. The Messerschmitt Bf 110Cs were not used in combat before December 1939.

On December 5, 1939 eight Bf 110C–2s of I./ZG 26 took off from the Jever to intercept a RAF bomber formation on a recce mission at the Helgoland. No combat followed. The next day 2./ZG 26 lost a Bf 110 (Fw. Franz Hoffmann and Uffz. Wilhelm Fuchs) which collided in the air with... a British Avro *Anson* (VX–R, serial K6189) of 206 Coastal Command Squadron. A Bf 110C of I./ZG 76 attacked by mistake a friendly Do 18 flying boat (8L+FK) of 2./KüFlGr 406. On December 14, Bf 110C–2s of 2./ZG 26 were involved in a combat with *Wellingtons* of 99 RAF Squadron which attacked German ships at the Jade estuary. British air gunners damaged Hptm. Restemeyer's plane (W.Nr. 1809).

The German High Command reinforced the I./ZG 26 Bf 110C–2s with an ad hoc established Jagdgeschwader 1 (also called Jagdgeschwader 'Schumacher', after the name of the C.O.). The unit's aircraft took part in a great air battle off the Helgoland. The German fighters encountered a large formation of *Wellingtons* from the No. 9, 37, 38, and 149 RAF Squads. The Bf 110 C–2s were patrolling the air spece over Frisian Islands. Pilots of 3./ZG 76 were the first to achieve kills. At 12:40 p.m. Fw. Groning shot down a *Wellington*, and he was soon followed by Olt. Gordon Gollob (later a leading Luftwaffe ace) and Ofw. Fleischmann. The second formation of British planes was intercepted by Bf 110Cs of 2./ZG 76 and and a 1./ZG76 plane flown by Lt. Helmut Lent. Lent shot down two aircraft of 38 Sqn (1 probably). One aircraft was destroyed by Olt. Jäger of 2./ZG 76. Other 2./ZG 76 pilots claimed four machines: two by Hptm. Wolfgang Falck and two by Uffz. Fresia. The table gives the score of ZG 76 Bf 110s.

Time	Pilot	Area where the kill was achieved
14.35	Hptm. Wolfgang Falck	20 km South–West of Heligoland
*14.35	Hptm. Wolfgang Falck	as above
14.35	Uffz. Heinz Fresia	as above
14.35	Uffz. Heinz Fresia	as above
14.42	Lt. Helmut Lent	5 km North of Borkum
14.50	Lt. Helmut Lent	10 km North–West of Borkum
14.45	Olt. Hans Jäger	North–West of Borkum
*14.45	Olt. Gordon Gollob	West of Langeoog
14.40	Fw. Hans Gröning	East of Langeoog
14.45	Ofw Georg Fleischmann	Spiekerooge
15.00	Olt. Walter Grenes	25 km North–West of Borkum
15.00	Lt. Maximilian Graeff	25 km North–West of Borkum
15.00	Uffz. Ernst Kalinowski	as above
15.00	Lt. Gustav Üllenbeck	50 km North of Ameland
15.00	Lt. Gustav Üllenbeck	as above

* kill not confirmed (probables), given according to German time.

On January 5, 1940 ZG 26 of Luftflotte 3 attained initial operational capability. Three days earlier a Schwarm of I./ZG 76 BF 110s shot down three 149 Sqn *Wellingtons* in a few minutes. On January 10, the Messerschmitts intercepted a formation of Bristol *Blenheim* bombers of 110 Sqn, RAF over Heligoland; only one British aircraft was shot down. On February 17, 1940 Hptm. Wolfgang Falck shot down a 110 Sqn machine during a recconnaissance flight over Heligoland. During February more units re–equipped with Bf 110Cs; I./ZG2 (ex–JGr 102),and III./ZG26 (ex–JGr 126). On March 26, 1940, Bf 110Cs of V.(Z)/LG2, flying escort to a reconnaissance Do 17 entered combat with *Hurricanes* of No. 73 Sqn RAF; three Messerschmitts were slightly damaged. On April 1 there was another combat, and British fighters damaged the aircraft of Ofw. Eugen Blass. The following day, at Luneville in Lorraine, a French Morane–Saulnier MS 406C–1 of GC II/3 shot down a Bf 110C of V.(Z)/LG1. On April 7 a formation of Bf 110Cs of I./ZG2, led by Hptm. Hannes Gentzen, fought again with the French; Gentzen shot down a Curtiss H–75C–1 (A–2?)

The Bf 110 C–2 of LG 1 flying in formation over Paris.
(MVT via M. Krzyżan)

Personnel of 8./ZG 76 during general Kesselring inspection, France 1940.
(Archiwum Dokumentacji Mechanicznej)

of GC I/5, while the French fighters claimed two Bf 110Cs. On May 9, 1940, a day before the attack on France, British pilots of No. 87 Sqn heavily damaged a Bf 110C of II./ZG1, which was forced to land.

During the Scandinavian campaign against Denmark and Norway, the following units of Bf 110Cs were involved: I./ZG76, led by Hptm. Reinecke, with 32 aircraft at the Westerland airfield, 3./ZG1 of Hptm. Ehle with 10 aircraft at Westerland, and I./ZG1 at Barth, under command of Hptm, Falck, with 22 Bf 110Cs.The first combat mission of Messerschmitt Bf 110Cs during the Operation Weserübung was the escorting of Junkers Ju 52/3m transports that carried the paratroopers who were to capture important military objectives; the 'Green Devils' were to take control of the airports of Oslo (Fornebu, also spelled Fornebeu), Stavanger–Sola and Aalborg in Denmark. Aalborg airport was the first target. Hptm. Wolfgang Falck of I./ZG1 shot down a Danish Fokker D.XXI during take off from Vaerlse airfield. The Messerschmitts destroyed ten Fokker C.VE planes of the 1 and 2 (reconnaissance) Eskadrille and four Fokker D.XXI fighters of the 2 (fighter) Eskadrille on the airfield. Lt. Schmidt, Olt. Martin, Olt. Lutz and Uffz. Maichi destroyed two aircraft each, while Olt. Streib, Olt. Müller, Lt. Wandam, Uffz. Dünsing and Uffz. Eberlein claimed one aircraft each. Two Bf 110Cs were slightly damaged by the AA artillery. Soon after the capture of Aalborg airport the aircraft of 1./ZG 1, 3./ZG 1 and 2./ZG 76 landed there. In the afternoon they were joined by the 2./ZG 1 aircraft. Transport planes brought spares and fuel. ZG 1 was charged with air defence of the airway to Oslo and Stavanger.

The action at Oslo was completely different. Because of fog, the first wave FJR 1 paratroopers were not dropped. In spite of that the leader of the second wave of transport aircraft decided to attack the Oslo–Fornebu airport. In the vicinity, the escorting Messerschmitts fought Norwegian Gloster *Gladiators*. One of the Norwegian fighters was shot down by the Bf 110C of Uffz. Mütscherle and Gefr. Lorey of 1./ZG 76, which force–landed moments later. At the airport, Olt. Hansen destroyed two *Gladiators*, while the machine of Lt. Erhard Kort and Uffz. Heinrich Bockheimer was destroyed, killing the crew. When the transport aircraft came in to land, Olt. Hansen in command with 1./ZG 76 gave order for the Bf 110Cs to attack the ground AA stations around the airport; this proved unnecessary, as the Norwegian soldiers had already left their posts. Several Bf110s landed at the edge of the runway and opened fire towards enemy positions at the other end; during the attack five aircraft were slightly damaged by machine gun fire.

At Stavanger the transports were protected by eight Bf 110Cs of 3./ZG 76. The aircraft of Uffz. Gerhard Grams and Lt. Sido Habben collided in the fog, both crews being killed. On April 9 at 6:00 p.m., two Bf 110s flown from Oslo–Fornebu by Olt. Hansen and Lt. Lent intercepted a Short *Sunderland* flying boat (L2167) of No. 210 Sqn RAF on a patrol flight; the *Sunderland* was shot at by both fighters, and Lt. Lent's fire proved deadly accurate blowing up the British aircraft in the air. Only the radio–operator was saved, having been thrown out of the plane without his parachute, he survived a fall from 500 m!

On April 10 the Bf 110Cs of 2. and 3./ZG 76 flew from Aalborg to Norway. On April 11 the Bf 110s from Stavanger–Sola tried to intercept British bombers that attacked their airfield, but with no result. On April 11, 2./ZG 76 aircraft were also transferred to Stavanger.

When the 'Battle of France' began, only two units equipped with Bf 110s were based in Denmark and Norway; these were 2./ZG 76 under Olt. Gustav Üllenebeck, with 11 aircraft (7 serviceable) at Aalborg in Denmark, and 1. and 3./ZG 76 under Hptm. Restemeyer with 25 (16 serviceable) Bf 110s at Stavanger–Sola. On May 13 Uffz. Herbert Jacobi of I./ZG 76 shot down a *Blenheim* at Bergen,and on May 16 two Bf 110Ds, equipped with additional fuel tanks, escorted He 111H bombers of KG 26 to Narvik. The next day I./ZG 76 aircraft shot down another *Blenheim* at Bergen. On May 20, 1940, the Bf 110Ds escorted bombers to Narvik again,while on May 23 the Bf 110s escorted Junkers Ju 52/3m transports, carrying the 2 Mountain Division's I./GJR 137 soldiers.The following day 3./ZG 76 lost a Bf 110, which crash–landed at Bardufos airfield after a combat with a No. 263 Sqn RAF *Gladiator*. On May 27, Lt. Helmut Lent damaged a *Gladiator* which force–landed. The Messerschmitts not only fought against aircraft, for on May 28 four Bf 110s of I./ZG 76 sank a small Norwegian vessel. On June 2, 1940 eight Bf 110Ds of I./ZG 76 flew missions over Narvik; the Messerschmitts attacked enemy ground forces. Lt. Lent of 1./ZG76 shot down the *Gladiator* II (N5893) of F/O J. L. Wilkie at Rombaksfiord. On June 11 a Bf 110D (M8+HH) of 3./ZG 76, flown by Lt. Kamp, shot down a *Hudson*. The night of June 13, 1940 was rather unlucky for the Fleet Air Arm pilots. Over Trondheim, the ZG 76 Messerschmitts intercepted a formation of No. 803 Sqn FAA Blackburn *Skua* dive–bombers, sent from HMS *Ark Royal* to destroy the *Scharnhorst*, the German battleship moored in that port. One of the British planes fell to Olt. Gollob, another to Lt. Krzywon (M8+BL), two more to Lt. Khler and Ofw. Schob. On June 15 Lt. Bender and Lt. Lent shot down a Bristol *Blenheim* IVF fighter of No. 254 Sqn RAF Coastal Command. The British plane fell on Smola island. On June 23, another *Blenheim* was shot

The pair of Bf 110 D of 7./NJG 4, France, 1942.
(P. Jarrett via B. Ketley)

The Bf 110 C–4 of EprGr 210 flying in formation – 1940, period of bomber raids over Britain.
(P. Jarrett via B. Ketley)

down by Lt. Bender of I./ZG 76. On July 9, 1940 the Messerschmitts shot down six Bristol *Blenheim* Mk IV bombers of 21 and 57 RAF Squads during their attack at Stavanger–Sola. The same day, at 2:20 p.m. Lt. Gordon Gollob claimed a Short *Sunderland* Mk I flying boat ('Y' N6133) of No. 201 RAF Coastal Command Squadron. At 5:20 p.m. the same pilot shot down a *Hudson* Mk II of 224 Squadron, west of Bergen. On August 15, 1940 the Bf 110Ds of I/ZG 76 flew escort to 63 He 111Hs of I. and III./KG 26 against targets in East Anglia. The RAf fighters shot down no less than seven Bf 110s. At the end of August the I./ZG 76 was withdrawn from Norway, superseded at Stavanger by the III./ZG 76. In 1941 the 13.(Z)/JG 5 was based in Norway.

BATTLE OF FRANCE – 1940

The following units equipped with Messerschmitt Bf 110Cs took part in the 'Battle of France': ZG 1, I./ZG 2, ZG 26, I./ZG 52, II./ZG 76, I.(Z)/LG 1 and LG 2. When the attack against France started, the Luftwaffe possessed 335 Messerschmitt Bf 110Cs. 238 aircraft were used operationally. On May 10, 1940 the Luftwaffe lost two Bf 110C–2s of II./ZG 26, shot down by the pilots of 1 RAF Squadron. The following day three Bf 110s fell at Rethel, hit by RAF pilots. The Bf 110 main task was to protect own bomber formations. The Messerschmitts were particularly useful when attacking targets far beyond the front line, exceeding the Bf 109E range. The escort to Dornier 17Z–2s of I./KG 76 attack against Hirson was an example of such a mission. The Messerschmits escorted bombers to Paris on June 5 (Unternehmen 'Paula'). The Bf 110s also flew sweeps (Freiejagd – free hunting) against British and French fighters and bombers. On May 18, when fighting the No. 73 RAF Squadron *Hurricanes*, the ZG 76 C.O., Oberst Grabmann was shot down.

The Bf 110 was much superior to its French counterpart, the Potez 631C2/CN2 or 63–11C3. It also bettered in both speed and armament the Curtiss H–75C1, Morane Saulnier MS.406C1 and Marcel Bloch MB.151 – 152C1 fighters. Only the new Dewoitine D.520C1 was faster.

During the Battle of France, the Luftwaffe lost 82 Bf 110s. Such a great number of losses resulted from its wide use; some 30% of the losses were for non–combat reasons. The Bf 110 suffered heavy losses from the British *Spitfires* and *Hurricanes* when attacking Allied forces during the evacuation from Dunkirk. The RAF fighters shot down no less than 43 aircraft of the type.

The reconnaissance unit Messerschmitts flew many missions; the 7.(F)/LG 2 flew such sorties over the British Isles from Grimbergen in Belgium. Several Bf 110s were shot down by Polish pilots in French units, as well the Polish GC I/145 squadron.

Basically the Messerschmitt Bf 110 proved successful in the Battle of France during May

The Bf 110 C–4 (A2+CL) of 2./ZG 52 over Dover, summer 1940, Battle of Britain.
(MVT via M. Krzyżan)

The Bf 110 D of NJG 3. Plane with designationof LG 1 (3C+BN). Under the cockpit on the left fuselage side was painted small flags of following countries: Belgium, France, Holland, Greece, Iraq, Yugoslavia, Finland and Denmark.

(P. Jarrett via B. Ketley)

and June 1940. The true test of the '110', however, was to come during the Battle of Britain.

OVER BRITAIN 1940 – 1943

Messerschmitt Bf 110Cs appeared over England for the first time in early July 1940. On July 6, 1940, in undisclosed circumstances, a Bf 110C–5 of I.(F)/Ob.d.L. went missing when flying a reconnaissance mission to Aberdeen; three days later, when attacking a convoy in the English Channel, the Luftwaffe lost two Bf 110s: one of III./ZG 26, and another of V./ZG 1. While escorting Dornier Do 17Z–2s of KG 2 which attacked the 'BREAD' convoy, *Spitfires* of 64 RAF Squadron shot down a Bf 110C–2 of 8./ZG 26 (3U+GS); the pilot, Ofw. Wolfgang Meyer, was taken prisoner, while the gunner, Gefr. Walter Rohde was killed. On July 11, 1940, 9./ZG 76 lost three aircraft over Portland (including one W. Nr 3531, 2N+EP) shot down by the British fighters; A nephew of Reichsmarschall Göring was killed in one of the planes. On July 21, 1940, during a reconnaissance flight over Southern England, the aircraft of No. 238 Sqn RAF hit a Bf 110C of 4.(F)/14. The Messerschmitt force – landed at Goodwood Home Farm near Chichester; the crew, pilot, Olt. R. K. Kunde and gunner, Fw. W. Baden, were taken prisoner. The aircraft, W.Nr. 2177, 5F+CM, was later repaired and flight tested.

On the 'Adlertag', the first day of mass attacks against airfields, the Luftwaffe lost several Bf 110Cs flying escort to bomber formations. At Bournemouth RAF fighters shot down several Bf 110Ds of LG 1; 13./LG 1 lost two machines, while 15./LG 1 lost one Bf 110D. When escorting KG 51 bombers, one Bf 110C–4 of 1./ZG 2 was shot down by Sgt. L. N. Guy of No. 601 Sqn RAF, while a 3./ZG 26 machine was claimed by P. S. Weaver of No. 56 Sqn. On August 14 at noon, the Erprobungsgruppe 210 (a unit formed for Messerschmitt Me 210 crew training) carried out a bombing attack against Manston airfield. The Bf 110s came at an altitude of about a twelve metres and successfully attacked the airfield. During the dive–bombing attack, the AA artillery shot down two Bf 110Ds (S9+MK and S9+NK) of 2./ErpGr 210, both crews being killed. On August 15 the machines of I./ZG 76 escorted the Heinkels of KG 26 during an attack against *Sunderland*. At 3:00 p. m., Martlesham Heath was the target for Bf 110Ds of ErpGr 210. In the evening, at 7:00 p. m., the same unit was directed against Kenley airfield, but they did not find the target and attacked South Croydon instead. Warned British fighters decimated the Messerschmitts. Among others, No. 32 and 111 Sqn RAF *Hurricanes* shot down the machine of the unit's C.O., Hptm. Walter Rubensdörfer (W.Nr. 3338, S9+AB). Also the Gruppenadjutant, Otl. Heinz Fiedler's aircraft (W.Nr. 3374, S9+BB) was shot down, as was that of engineering officer, Lt. Ludwig Koch. Other dive–bombing elements of ErpGr 210 also suffered losses. Among others, the Bf 110C–6, S9+TH of 1./ErpGr 210 was lost, as was the S9+CK of 2./ErpGr 210, the crew of which (Olt. A. Habisch

The Bf 110 C–4 (A2+CL) of 2./ZG 52 on the airfield in France, Battle of Britain, summer 1940.

(MVT via M. Krzyżan)

The pair of Bf 110D of III./ZG 26 in flight, 1941.
(MVT via M. Krzyżan)

and Uffz. E. Elfner) became PoW's. A Bf 109E–4 equipped experimentally with an underfuselage bomb rack was also lost. On the same day, the 1. and 2./ZG 76 aircraft escorted KG 30 Ju 88s against the East Anglia airfields.

One of the missions of August 15, 1940 is recalled by Uffz. Max Guschewski of ZG 76:

'In the morning we were moved from le Mans to Rennes. There the Bf 110s were fitted with extra fuel tanks. We took off on August 15 after 4:00 p. m. Over Guernsey we took escort of 40 Ju 88s. (...) At 4,000 m altitude we passed Cherbourg, going towards England. Over the Channel we climbed to 4,000 m. The Salisbury airfield was our target. The sky above the Isle of Wight and Portland was clear, but for the condensation trails forming behind our aircraft. Over the target we were attacked by RAF Spitfires. A combat started. The bombers, upon dropping their bombs, turned for France. Over the Solent our aircraft was attacked by four British fighters. Two of them attacked head–on, two from the tail. The starboard engine was hit and went to flames, the port one stopped, too. When I was trying to abandon the aircraft, I was hit with a bullet, luckily the pilot managed to land in a small field. After the landing I became PoW.'

Guschewski's aircrft was shot down by two Polish fighters – Piotr Ostoja–Ostaszewski of No. 609 Sqn RAF and Janusz Żurakowski of No. 234 Sqn. The Messerschmitt Bf 110C–2, M8+BP, of 6./ZG 76 landed at North Ashley Down at 6:06 p. m. The pilot, Fw. Johann Binsdorfer, was shot while attempting to escape, while the wounded gunner, Uffz. Max Guschewski was taken prisoner. Thefollowing day the Luftwaffe lost, among others, a Bf 110C of 6./ZG 2 shot down over Eastbourne. August 18, was a black day for the ZG 26 which lost no less than eleven Bf 110Cs, including U8+BB, flown by the I./ZG 26 Gruppenadjutant, Olt. R. Proske. Uffz. H. Jäckel was killed in 3U+AM of 4./ZG 26. 6./ZG 26 suffered particularly heavy losses, four aircraft of this unit being shot down. The 1./ZG 2 commander was also shot down on August 25.

On August 30, 1940 F/O Ludwik Paszkiewicz of No. 303 (Polish) Sqn RAF shot down the Bf 110C–4, (W.Nr. 3615; M8+MM) of 4./ZG 76, flown by Ofw. Georg Anthony. The following day British fighters destroyed two Bf 110Cs of 14./LG 1 over the Thames Estuary. On September 1, 1940 Bf 110s of 13. and 15./LG 1 attacked Biggin Hill airfield. The next day the Luftwaffe lost, among others, the Bf 110C–4, 3M+HK (W.Nr. 3622) of I./ZG 2. The crew, Lt. Schipper and Hptm. Schockendort became PoW's. The same day brought destruction of the Bf 110C–2 W.Nr. 3226, M8+DM of 4./ZG 76 and death of the crew – Olt. Wreder and Uffz. Kuklewski.

On September 7, the following units took part in actions over Britain (see table).

Table
Units, who took part in action over Britain on September 7, 1940

Unit	Number of A/C	Serviceable A/C	Airfield
Reconaissance and bomber units			
Luftflotte 2			
1.(F)/22	7	?	Lille
4.(F)/122	6	?	Brussels
Fliegerkorps II			
EprGr 210 (kpt. Lutz)	26	17	Demain/
7.(F)/LG 2			Valenciennes
Luftflotte 3	14	9	–
Fliegerkorps IV			
3.(F)/31	9	5	Saint-Brieuc
4.(F)/14	12	9	Normandy
Fighter units			
Luftflotte 2			
Jagdfliegerkorps II			
ZG 26 (Otl. Hüth)			Lille
I./ZG 26 (Hptm. Makrocki)		19	Yverench
II./ZG 26 (Hptm. von Rettburg)		–	Crecy-en-Ponthieu
III./ZG 26 (Hptm. Schalk)		16	Barley
ZG 76 (Maj. Grabmann)			Laval
II./ZG 76 (Hptm. Groth)		–	Abbeville–Drucat
III./ZG 76 (Hptm. Dicrode)		22	Laval
Fliegerkorps VIII			
V.(Z)/LG 1 (Hptm. Liesenberger)		–	Caen–Carpiquet
Jagdfliegerkorps III			
ZG 2 (Otl. Vollbracht)			Toussoun
I./ZG 2 (Hptm. Heinlein)		16	Amies
II./ZG 2 (Maj. Carl)		–	Guyancourt
Luftflotte 5			
I./ZG 76		9	Stavanger–Sola

On September 9, 1940, 9./ZG 76 lost two Messerschmitt Bf 110s over the English Channel. Two days later count Czernin an ace of No. 17 Sqn RAF shot down over Thaned a Bf 110C–4 of Stab/ZG 26 (W.Nr.3625, 3U+HM), flown by Olt. Werner Henken. On September 25, 1940, a Bf 110C–5 (W.Nr.2185, L2+ER) of 7.(F)/LG 2 was shot down during a reconnaissance flight over London, and the following day the same happened to a plane of 4.(F)/14 (5F+CM). On September 27, 1940, the Messerschmitts of ErpGr 210 carried out a dive–bombing attack against Parnall aircraft company at Bristol. Near the target they were intercepted by *Hurricanes* of No. 504 RAF Sqn that shot down two Bf 110D–3s of Stab and 1./ErpGr 210.

September 27, proved in fact a slaughter day for the Bf 110–equipped units. No less than 18

The Bf 110 D–3 of ZG 26 flying in formation.
(MVT via M. Krzyżan)

Commander of ZG 26, Lt. Schalk getting off Bf 110 cockpit. Note unusual function markig on the fuselage.

(MVT via M. Krzyżan)

aircraft of the type were shot down that day. One of the losses was Hptm. Horst Liensenberger, the C.O. of V.(Z)/LG 1 who was killed in his plane. P/O Percy Burton of No. 249 RAF Sqn, after having used up all the ammunition, rammed Liensenberger's aircraft! Seven aircraft out of 13 taking part in that mission did not return. The decimated unit was withdrawn from operations and relocated to Bavaria, where it formed part of the newly organized NJG 1.

Losses of V.(Z)/LG 1 on September 27, 1940

Aircraft Number, Type Unit Code	Crew	Area Time by whom shot down Unit
Bf 110 C–2 (W.Nr. 3560) Stab. V./LG 1 L1+XB	Hptm. Liesen-berger+ Uffz. Köpge+	Hailsham 09.50 P/O Burton 249 Sq. RAF
Bf 110 D–0 (W.Nr. 3333) 13./LG 1 L1+BH	Hptm. Swetlik+ Gefr. Weltz+	English Channel 09.40 Sgt Wright 605.Sq. RAF
Bf 110 D–0 (W.Nr. 3304) 13./LG 1 L1+CH	Fw. Bruns+ Fw. Gröbl+	Chelwood Gate 09.40 213. 605. Sq. RAF and 1 Sq. RCAF
Bf 110 C–2 (W.Nr. 3548) 14./LG 1 L1+CK	OGfr. Hübner+ Fw. Lindemann+	Coppice Farm 09.40 pilots of nos 249 and 253. Sqns RAF
Bf 110 D–0 (W.Nr. 3147) 15./LG 1 L1+BL	Olt. von Grafen-reuth+ Fw. Reinhold+	Gatwick 10.00 pilots of nos 17, 213 and 602 Sqns RAF and 1 Sqn RCAF
Bf 110 C–2 (W.Nr. 3849) 15./LG 1 L1+GL	Uffz. Bechtold Uffz. Koch+	Horsham Manor Farm 09.50 pilots of nos 17,46, 303,*(P) 501 and 602 Sqns RAF
Bf 110 C–2 (W.Nr. 3533) 15./LG 1	Olt. Weckeiser Gefr. Brüsgow	Socketts Manor 09.50 pilots of nos 17, 46,92 249** i 605 Sqns RAF

Polish pilots:
* W. Urbanowicz of No. 303 (Polish) Sqn,RAF
** J. Budziński of No.249 Sqn,RAF

On October 7, 1940 a new Messerschmitt Bf 110E–1 assigned to the Stab.III./ZG 26 fell victim to RAF fighters. That day the Messerschmitts flew escort to II./KG 51 bombers attacking Yeovil. 9./ZG 26 lost three aircraft, while 4. and 6./ZG 26 an aircraft each. The latter lost a rarely used Bf 110C–7 (W.Nr. 3418, 3U+JP); this was a fighter–bomber fitted for the carriage of 500 kg bombs. On November 17, 1940, the 1./ErpGr 210 attacked Aldeburgh with bombs. On December 8 RAF fighters shot down a recce Bf 110C–5 (W.Nr. 2256, 5F+DM).

After the Battle of Britain was over, the Bf 110–equipped units were gradually withdrawn from the French bases to the Reich. On February 20, 1941, when attacking Harwich, a Bf 110E–1 of 3./ErpGr 210 was shot down by the HMS *Bramble* minesweeper's guns. During March of 1941 the aircraft of the unit, while taking off from the Belgian airfield at Ursel, carried out several night bombing attacks against targets in East Anglia. That was a completely new use for the Bf 110.

On March 11, 1941, in unsolved circumstances, a Bf 110E–3 of 3.(F)/123 (W.Nr. 2316, 4U+XL) went missing over the English Channel. Also 4U+SL (W.Nr. 2309) was shot down by a *Spitfire* of No. 234 Sqn RAF. During the same month British artillery destroyed a Bf 110E–2 (W.Nr. 3774) of Stab./ZG 76.

In April 1941 some elements of ErpGr 210 were used to form Schnellkampfgeschwader 210. This was led by Maj. Walter Storp. On May 19 1./SKG 210 lost a Bf 110E–1 (W.Nr. 3483, S9+BL), shot down by AA artillery over Lowestoft. On August 10, 1941, 1.(F)/123 lost a Bf 110 over the Channel; the same unit lost a Bf 110C–5 (W.Nr. 2237, 4U+YL), shot down by a *Hurricane* IIC of No. 87 Sqn RAF on October 21, 1941. On July 21, 1942, 3.(F)/123 lost a Bf 110E–3 (W.Nr. 2356, 4U+ZL) shot down by a *Spitfire* of 234 Sqn.

The last Bf 110 to be lost over Britain was a Messerschmitt Bf 110G–2 (W.Nr. 150130) of 4./ZG 1, shot down over the Scilly Isles by *Spitfire* VBs and IXs of Nos 453 RAF and 610 AAF Sqns.

The Battle of Britain fighting finally destroyed the myth of the Messerschmitt Bf 110. When encountering strong fighter defence the Bf 110s had to be escorted by the Bf 109s. The defensive armament, a single MG 15, was totally insufficient. Also the concept of using the Bf 110 as a fast bomber and dive bomber failed completely. The Messerschmitts suffered No. less losses than the Dornier and Heinkel bombers.

Also the tactics of the so called defensive ring proved inefficient; when the formation broke loose after being attacked, the remaining aircraft fell easy prey to the enemy fighters.

The Bf 110 B–3 of training unit, Sicily, 1941.

(P. Jarrett via B. Ketley)

The Messerschmitt Bf 110 D–3 of ZG 76, Balkans, 1941. *(MVT via M. Krzyżan)*

We should not forget, however, the successes of the Bf 110s. The well armed Bf 110 (two MG FF cannon and four MG 17) could be a dangerous opponent once attacking unnoticed or bouncing from high above. During the Battle of Britain several pilots gained much success, like Sophiuss Baagoe of 7./ZG 26 (9 kills), Erich Groth of 8./ZG 2 (12 kills), Richard Heller of 9./ZG 26, and Heinz Nacke.

BALKANS, CRETE – 1941

Several Messerschmitt Bf 110 units were assigned to operation 'Marita', the attack against Yugoslavia and Greece. I./ZG 26, based at Szeged in Southern Hungary, flew within the Luftflotte 4; II./ZG 26, based at Krainitsi and Sofia–Vrazhdebna in Bulgaria, was assigned to the VIII Fliegerkorps. The reconnaissance Bf 110C–5s of 7.(F)/LG 2 were based also at the latter airfield.

The Messerschmitts participated in the attacks against the Greek, Yugoslav and the Greece–based British air forces from the first moments of the conflict. In the morning of April 6 they took part in the mass attacks against Yugoslavia.

It is interesting to note also that the Royal Yugoslav Air Force had on strength a single Bf 110! On March 27 (or 29) 1941, a Bf 110D of 7.(F)/LG 2 landed in error at Kraljevo airfield during a reconnaissance mission. It was interned and impressed into the No. 51 Fighter Group at Zemun near Belgrade. The fate of the machine during the April 1941 campaign remains unknown.

During the attack against Rezovnacka Kosa a single Bf 110D of LG 2 took part in combat against Hawker *Furies* of No. 36 Group. F/L Konstantin Ermakov of No. 112 Eskadrila, 36 Group, tried to ram a Bf 110. II/ZG 26 lost two Bf 110s; in one of these a Bulgarian officer, who was guiding the Luftwaffe formation to the target, was killed. At Petrovac airfield the Bf 110s of II./ZG 26 destroyed 14 Yugoslav Dornier Do 17K bombers of No. 54 Group on the ground.

Yugoslav AA artillery shot down a Bf 110 of II./ZG 26. On April 6, in the afternoon, the Bf 109s and Bf 110s 'grounded' the remaining Dornier Do 17Ks of No. 63 Group at Petrovac, while that morning the Bf 110s of I./ZG 26 had flown escort to the German II./KG 4, KG 2 and KG 3 against Belgrade and escorted other aircraft at medium level (3,000–4,000 m). During a combat against Yugoslav Bf 109Es (Messerschmitt vs. Messerschmitt!), I./ZG 26 claimed several Bf 109s. Two of these fell to Uffz. Stiegler. That day the I./ZG 26 lost five Bf 110s.

In the afternoon III./ZG 26 aircraft based in Italy attacked the Mostar airfields (Podgorica and Mostar–Ortijes). The Messerschmitts destroyed two Yugoslav Savoia–Marchetti S.79 bombers on the ground and Ofw. Richard Heller of III./ZG 26 damaged a Yugoslav Bf 109 south of Sarajevo.

On April 7, the Bf 110s attacked Lazarevac, where they destroyed 12 training aircraft on the ground. In the afternoon, when escorting bombers, a Bf 110 of I./ZG 26 was shot down by a Yugoslav Ikarus IK–Z fighter,and II./ZG 26 lost a Bf 110 which crashed during a forced–landing south of Sofia.

On April 11, the Messerschmitts of I./ZG 26 attacked the Voivodina airfields. At Bjeljina a section of Bf 110s destroyed two Bristol *Blenheim* Is of No. 7 Regiment, several Breguet XIXs and training Bücker Bü 131 *Jungmanns*.

At 7:30 the following morning, the Bf 110s recommenced their attacks, destroying a record number of aircraft. The Yugoslav Air Force lost no less than 17 *Blenheims* (10 of No. 1 Regiment and 7 of No. 11 Group), several Breguet XIXs, almost all the Hawker *Hurricanes* of No. 105 Eskadrilla, and some training aircraft. Only two *Blenheims* and a *Hurricane* were left serviceable after the attack.

On April 14, 23 Bf 110 fighters of I./ZG 26 (Fliegerführer Arad) provided air cover for the landing of Ju 52 transports at Bjeljina airfield, where they delivered troops to capture the place.

On April 20 the Bf 110 of II./ZG 26 attacked Menidi, destroying several RAF aircraft, as well as a Greek Junkers (G 24 or Ju 52/3m). A formation of German aircraft was led by Olt. Ralph von Rettburg. During an afternoon attack against Athens, the Messerschmitts escorted a formation of some one hundred Ju 88 and Do 17 bombers. The Bf 110s not only provided escort but also tried, without great results, to attack the *Hurricanes* of Nos. 33 and 80 Sqn RAF taking off from Eleusis at Athens. The leading RAF fighter ace of the WWII, S/L Marmaduke Thomas Saint John Pattle of 33 Sqn, credited with 50 2/3 kills, was killed in the combat with the Messerschmitts. Another ace, F/L Vernon Woodward, shot down a Bf 110, while another Messerschmitt fell to Jimmy Kettlewell. Sgt Cottingham's *Hurricane* was shot down in return. The German pilots claimed five victories. These were credited to Hptm. Theodor Rosinwall (12th kill), Olt. Sophus Baagoe (14th), Fw. Theodor Pietschmann, Hermann Schönthier and Uffz. Fritz Müller.

On April 21, 1941, Messerschmitt Bf 110s of 5./ZG 26 again attacked Eleusis; two aircraft collided in mid–air over the target, but they were only slightly damaged and returned safely to base. In the evening of April 23, 40 Bf 110s of I./ZG 26 led by Gruppenkommandeur Maj. Wilhelm von Makrocki attacked Argos airfield, achieving hits on all the *Hurricanes* and Greek

The Meserschmitt Bf 110 D–3s of 4./ZG 76 of the airfield in Greece. Note white covers of the nose armament and white spinner tips – elements of rapid identification system during that campaign. *(MVT via M. Krzyżan)*

The pair of Messerschmitt Bf 110 D of III./ZG 26, Greece, 1941. *(MVT via M. Krzyżan)*

training aircraft left there. The Bf 110s destroyed more than 40 aircraft, including 13 *Hurricanes* and a No. 208 Sqn *Lysander*, plus some Greek aircraft. During the attack against Menidi airfield, the AA artillery shot down an aircraft of 6./ZG 26, killing the crew of Uffz. Kurt Schwamm.

On April 26, 1941 the German paratrooper units attacked the Corinthian Canal; the ZG 26 Messerschmitts destroyed the positions of British AA artillery placed around the Canal. One aircraft (U8+AL) of 3./ZG 26, flown by Uffz. Wilhelm Rödel was lost to AA fire, the crew being killed. The following day the I./ZG 26 aircraft attacked the retreating Allied troops on the roads. That day the Luftwaffe lost a single Bf 110 shot down over Athens, the crew escaping with wounds. During the Yugoslav and Greek campaigns, the Luftwaffe lost 25 Messerschmitt Bf 110Cs and Ds.

On May 11, 1941, the Heraklion airfield at Crete was the target of an attack by five Messerschmitt Bf 110Ds of II./ZG 76. The unit was transferred to Greece from Germany on May 4, 1941. Only F/O Bowker of No. 112 Sqn RAF took off in defence in a *Gladiator*.

The combat brought no result. The Allied pilot success was that he was not shot down by the five Bf 110s. On May 14, the leading II./ZG 26 pilot, Olt. Baagoe, was killed, being shot down by the Bofors guns defending the Heraklion airfield. Another Bf 110 (3U+EM) was also claimed bythe Royal Artillery. The crew of the 4./ZG 26 aircraft, pilot Gefr. Adolf Ketterer and gunner Gefr. Hans Bomba became PoWs. The Messerschmitts destroyed a *Gladiator* of No. 1430 Flight flown by Stan Reeves, and four other biplanes were damaged on the ground. On May 16, the Messerschmitts escorted Ju 87R–2s of I./StG 2 when attacking Maleme and Suda Bay. On the other hand the I. and II./ZG 26 aircraft attacked Heraklion airfield. During combat over the airfield F/L Fry of No. 112 Sqn damaged a Bf 110 (3U+SM) of 4./ZG 26, but the Messerschmitt's gunner, Fw. Karl Reinhardt, managed to shoot down the enemy fighter. On May 17 British *Beaufighters* of No. 252 Sqn RAF destroyed three Bf 110s and damaged 13 more aircraft at Argos airfield.

On May 18, using a BREN machine gun, Sgt. T. H. Hawksworth and Private R. G. Kimber of the Royal Marines shot down Lt. Heinz Knecht's Bf 110D of 6./ZG 26 which crashed into the sea. The crew was rescued and taken prisoners. Similar was the fate of a 4./ZG 26 aircraft, shot down by a vintage Lewis gun.

The Messerschmitt Bf 110s of I. and II./ZG 26, led by Maj. Johann Schalk, took part in operation 'Merkur' (the attack against Crete), accompanied by the reconnaissance machines of 7.(F)/LG 2 and the XI Fliegerkorps Staffel. On May 20, during an attack against Canea, Hptm. Paul Kleinert's aircraft (3U+HM) of 4./ZG 26 was shot down. Two Bf 110s; 3U+AM of Olt. Reinhold Heubel, the Staffelkapitn, and 3U+CN of Uffz. Otto Stein, were shot down during the afternoon attack against Heraklion. Both aircraft were from 4./ZG 26. The following day another machine was lost over Heraklion, this time a Bf 110 (3U+LN) of Lt. Dietrich Oldenburg of 5./ZG 26. The same day I./ZG 26 aircraft attacked the British battleship, HMS *Warspite*; the Luftwaffe lost an aircraft (3U+CC) of the Stab Staffel, flown by Olt. Erich Kissel. During the following days the Bf 110s also attacked the airfields on the island. On May 24, a Bf 110D–3 (M8+AC) of II./ZG 76 was shot down by a British machine gun mounted on the ML 1001 rescue boat. On May 28, 1941 the Bf 110s of II./ZG 26 and II./ZG 76 escorted KG 2 Do 17Zs when attacking Heraklion. The following day the Luftwaffe lost a II./ZG 76 aircraft, shot down by AA fire over Suda.

The losses not only resulted from enemy action. A Bf 110C–4 (3U+CP) of II./ZG 26 crashed into the sea because of an engine malfunction (although some sources cite it as a No. 45 Sqn aircraft kill).

AFRICA, MALTA – 1941–1943

The Messerschmitt Bf 110s were also used over Malta and Northern Africa.

On February 17, 1941, 1./NJG 3 Bf 110Ds appeared above Malta. The same day one aircraft was shot down by F/O Hamilton. Also on March 10, 1941 1./NJG 3 lost a Bf 110C over the island. On April 19, 1941 a damaged aircraft ditched near the island and the crew became PoWs. Sicily–based aircraft took part in the day attacks, on May 8, 1941 9./ZG 26 Bf 110s, led by Hptm. Thomas Steinberger escorted the I./StG 1 Stukas. In the combat with Nos 807 and 808 Sqn FAA *Fulmars* from HMS *Ark Royal,* the Messerschmitts slightly damaged two of the naval fighters. The Malta–based fighters were also credited with Luftwaffe aircraft operating in Africa. On December 22, 1941, III./ZG 26 lost a Bf 110 over Benghazi in this way. On January 16, 1942, a III./ZG 26 aircraft flown by Fw. Günther Wegmann shot down a *Wellington*. On February 12 an ace of III./ZG 26, Lt. Alfred Wehmeyer, shot down a *Beaufighter* (T4915) of No. 252 Sqn RAF; this was his tenth air victory.

The Messerschmitt Bf 110s were used not only as fighters; on March 21, 1942 eight III./ZG 26 aircraft attacked Malta's Hal Far airfield with bombs. On the night of April 12/13, three bomber Bf 110Es of III./ZG 26, escorted by Bf 109Gs, attacked Takali. On April 14, the Messerschmitts claimed two torpedo carrying Bristol *Beauforts*. On September 6, 1942, a Bf 110E of Fw. Wegmann, taking off from Kastelli in Crete, attacked a No. 159 Sqn RAF Consolidated *Liberator* II; the next day Wegmann shot down a Liberator and damaged another. His Messersch-

The pair of Messerschmitt Bf 110 D–3 of III./ZG 26 in flight over North Africa coast, summer 1941. *(MVT via M. Krzyżan)*

The Messerschmitt Bf 110 D–3 of III./ZG 26, Mediterranean Sea, summer 1941.
(P. Jarrett via B. Ketley)

mitt, 3U+AD, returned to Kastelli with 50 holes. On October 8, 1942, a formation of No. 160 Sqn RAF *Liberators* was attacked by Bf 110s of III./ZG 26. The crew of F/O Dupleix claimed two Messerschmitts; one damaged *Liberator* crashed on landing. On October 11, III./ZG 26 lost an aircraft in combat with Allied bombers, and on October 19 No. 272 RAF Sqn *Beaufighters* shot down a Bf 110 which escorted transport aircraft flying from Crete to Africa.

The first Bf 110s arrived in Africa at the end of January 1941, when III. Gruppe of ZG 26 was sent to Libya. The Gruppe was reinforced with 2. Staffel of I. Gruppe ZG 26 'Horst Wessel'; elements were stationed at Castel Benito, Syrta and Arco Philenorum. Maj. Karl Kaschka was the III./ZG 26 Gruppenkommandeur, and Olt. Fritz Schutze the Gruppenadjutant; Olt. Matthes was the 7./ZG 26 Staffelkapitän followed at the post by Olt. Georg Christl, 8./ZG 26 was led by Olt. Pang and later by Olt. Schutze – Dickow, and Hptm. Bord was the 9./ZG 26 Staffelkapitän. The units were equipped with Bf 110C–4s and D–3s.

On February 6 the unit suffered its first loss. The Bf 110 flown by Uffz. Lippki had to land in British–held territory because of an engine malfunction, and the crew was captured. On February 16, the machine of Lt. Hesterkamp and Uffz. Bracht was shot down by AA fire, the crew being killed. Three days later Messerschmitts shot down a *Hurricane* I Trop flown by P/O Gatward of No. 3 Sqn RAAF. F/L Perrin claimed a Bf 110, but was soon hit by another Messerschmitt and had to land in the desert. P/O Storrar of No. 73 RAF Sqn damaged a Bf 110 over Benghazi, inflicting wounds on the gunner, Gefr. Wust. Two more Bf 110s were lost to ground fire during February.

On March 29, 1941, Bf 110s attacked a No. 230 RAF Sqn *Sunderland* MkI, but without results. Three days later No. 3 RAAF Sqn shot down the Bf 110 of Fw. Josef Braun and Uffz. Max Kasper (gunner). On April 3, the III./ZG 26 lost no less than four Bf 110s shot down by P/O Turnbull of No. 3 RAAF Sqn. The Messerschmitts destroyed three recce *Hurricane* MkIs of No. 6 RAF Sqn. Two days later the Messerschmitts attacked British airfields destroying two Bristol *Blenheims* and a Westland *Lysander* MkII. On April 8, the Bf 110s of III./ZG 26 shot down a Bristol *Blenheim* MkIV of No. 55 RAF Sqn. On April 9, a Bf 110D–3 flown by Fw. Helmut Jacula was shot down by P/O Goodman of No. 73 RAF Sqn after a prolonged fight. The pilot was killed, while the gunner, Uffz. Wala, baled out wounded. In the afternoon the 7./ZG 26 C.O. – Olt. Christl shot down the *Hurricane* of Sgt. Elsworth of No. 73 Sqn RAF. The III./ZG 26 lost also an aircraft which crashed during a force landing at Tobruk, and the recce 2.(F)/14 lost a Bf 110C–5 shot down by the AA fire. Uffz. Heinz Straten, the pilot, was killed. On April 11, Bf 110s destroyed a No. 6 RAF Sqn plane. On April 14, Bf 110s took part in attack against Tobruk. They shot down Sgt. Webster's plane of No. 73 RAF Sqn over the fortress. On April 26, a No. 274 Sqn RAF aircraft and a Bf 110 of III./ZG 26 collided in mid–air. The Messerschmitt's crew. Lt. Oskar Lemcke and Uffz. Petters, were killed. Another No. 274 Sqn pilot, P/O Patterson, was shot down by a Bf 110. On April 30, III./ZG 26 lost another Bf 110. On May 20, 1941, during an allied attack against Mechili airfield, F/L Honor of No. 274 Sqn destroyed a Bf 110 taking off. On June 11, the 2.(H)/14 lost a Bf 110C–5 (Uffz. Otto Unger and Lt. Friedrich Giessalmann), shot down at Halfaya pass. On June 30, III./ZG 26 lost a Bf 110 claimed by P/O Hofe of No. 274 Sqn; another Bf 110 was damaged. On July 10, in the morning, the Bf 110s of III./ZG 26 attacked the Bagush satellite airfield, destroying a No. 113 Sqn RAF Bristol *Blenheim* IV on the ground. On July 15, two more Bf 110s were lost over Tobruk. Fw. Richard Heller of III./ZG 26 shot down a *Hurricane*. On August 11 a patrol of Curtiss P–40C *Tomahawks* of No. 2 Sqn SAAF claimed a Bf 110 of III./ZG 26. The crew of Uffz. Werner Seiler and Gefr. Schmitz was killed. During the combat over Tobruk the Bf 110s claimed four *Hurricanes*, losing only a single Messerschmitt. The Germans used the old technique, the defensive ring, which rendered successful attack by Nos 1 and 2 Sqn SAAF aircraft impossible. On August 29, 1941 the Messerschmitts scored a Martin *Maryland* II of No. 39 Sqn RAF. On September 3, the Germans destroyed a reconnaissance *Hurricane* of No. 451 Sqn RAAF, intercepted over Sollum. On November 15, 1941, when attacking airfields, III./ZG 26 lost three Messerschmitt Bf 110D–3s shot down by the AA fire; the 9./ZG 26 Staffelkapitän, Olt. Schultze – Dickow was shot down during that combat, and despite his wounds the pilot managed to land his machine on the bottom of a salt lake. The crew of 3U+DS was rescued by the Bf 110 of Fw. Swoboda, who landed beside the wreck and took them on board. In the afternoon of November 24, Allied aircraft, escorting motorized columns at Fort Maddalena, claimed four Bf 110s of III./ZG 26; in return the Germans shot down two No. 4 Sqn SAAF machines.

The Bf 110D of 9./ZG 26 during preparation to flight, Northern Africa, 1941.
(P. Jarrett via B. Ketley)

The combat is described by Richard Heller's gunner of 8./ZG 26: *'The sun was about to set when our Bf 110s arrived over the British tank columns. Our Rotte commander spotted a new column of trucks. Suddenly I spotted one, two, three, four small points. Fighters! — I shouted in my microphone. My pilot performed a sharp turn. Suddenly I saw blue—white—red roundels over my cockpit. The sky above us was full of enemy fighters. We tried to escape our pursuers by a shallow dive. The Curtiss was high above us, though, and he quickly entered a firing position behind our tail. During his first attack he hit our port engine. The Curtiss turned back quickly, but three more fighters appeared on our both sides. One Curtiss was circling above. The rounds were showering on us. We could not escape, both engines were out, and the starboard one trailed smoke. We only had one way left — to land in the desert. A Curtiss attacked us even after we landed, the pilot was slightly wounded in his hand.'*

On November 25 the Luftwaffe lost two Bf 110s of III./ZG 26 and another from the 2.(H)/14. During a sweep over Sidi Rezegh Bf 110s encountered Allied fighters during which two more machines were lost. The crews: pilots Olt. Dieter Bidlingmaier and Lt. Jorn Scharf, and gunners Uffz. Becker and Uffz. Bogler, were posted missing. The following day during a reconnaissance flight a Bf 110 of 2.(H)/14 was lost. On November 30, No. 238 Sqn RAF aircraft damaged a Bf 110E of III./ZG 26. On December 3, 1941 Sgt Dodds of No. 274 Sqn RAF probably destroyed a Bf 110. The British pilot did not see the result of the combat since a round from the German gunner's MG broke his goggles. The same day the aircraft of Maj. Karl Kaschka, the III./ZG 26 Gruppenkommandeur was lost, so it is most probable that this was the aircraft claimed by Dodds. Olt. Wehmeyer landed next to his C.O.'s machine, but the rescue came too late, the crew of Maj. Kaschka and Uffz. Muhlhausen was killed.

In December 1941 the Messerschmitts shot down a medevac de Havilland DH 86. This was not a singular case, as both sides of the conflict were attacking such aircraft, and RAF fighters shot down several Italian medevac planes. That day III./ZG 26 lost three aircraft, and one of 7./ZG 26 crashed at Derna when landing.

On December 9, 2.(H)/14 lost a Bf 110, the pilot Fw. Gnther Ursinus being killed and the gunner becoming a PoW. The following day Lt. Bromwick No. 805 Sqn FAA shot down a Bf 110 north of Ras de Kanayas. On December 11 Bf 110s encountered Allied fighters over Gazala, and on December 12, when providing air cover for a convoy, Lt. Schultze—Dickow shot down two *Marylands* of No. 21 Sqn SAAF, while Uffz. Wegmann claimed another South African machine. On December 16, seventeen aircraft of Nos 2 and 4 Sqns. SAAF met a reconnaissance Bf 110 over Gazala; they literally shot it out of the sky. On December 22, 1941 a Bf 110 of III./ZG 26 shot down a No. 230 Sqn RAF *Sunderland* between Egypt and Malta. On December 24, 1941 Hptm. Thomas Steinberger, the III./ZG 26 Gruppenkommandeur was killed during a ferry flight from Athens to Crete. Together with him Fw. Max Hoffmann a pilot of 9./ZG 26 was killed.

On January 5, 1942 over Benghazi, pilots of Nos 33 and 274 Sqns RAF damaged two Bf 110s. In mid–January only 7./ZG 26 with 8 aircraft (including 4 serviceable) was based in Africa. On January 10, 1942 the Messerschmitt Bf 110E–1/Trop, 5F+SK, of 2.(H)/14 was damaged during a forced landing. Maj. Heymeyer's Staffel was reporting directly to the German Afrika Korps Commander, Lt. Gen. Rommel. On January 22, 1942 *Tomahawks* of No. 250 Sqn RAF shot down a Bf 110 of 2.(H)/14 at Mersa Brega. The pilot, Olt. Freudenberger, and the gunner, became PoWs. Three days later Sgt Reid and Sgt Curtis of No. 3 Sqn RAAF shot down a Bf 110. On January 28 a III./ZG 26 aircraft crashed at Ghemines and the wounded pilot, Fw. Franz Sander, died on February 4, 1942. On February 8 and 10 Bf 110s attacked enemy armoured columns with bombs. Olt. Ludwig Schott's aircraft was destroyed by its own bomb dropped from too low altitude.

On April 5, 1942 Nos 94 and 260 RAF and No. 4 SAAF Sqns aircraft attacked Bf 109s of II./JG 27 escorting a reconnaissance Bf 110. On April 14, S/L Riley of No. 272 Sqn RAF (in a Bristol *Beaufighter* MkIC) shot down two Bf 110s that attacked torpedo–carrying Bristol *Beaufort* IIs of Nos 22 and 39 Sqn RAF. On May 9, 1942 while escorting Douglas *Boston* III bombers, S/L Gibbes of No. 3 Sqn RAAF damaged a Bf 110.

In May 1942 the Luftwaffe in the Mediterranean possessed the following Messerschmitt Bf 110 units:

— 9./ZG 26 at Maleme (Crete), 11 Bf 110Ds and Es

— 7./ZG 26 at Derna, 12 Bf 110Ds and Es.

On May 12, while escorting Junkers Ju 52/3m transprts of TG1, the Bf 110D–3 of Uffz. Baumann of 9./ZG 26 was shot down. Nos 250 and

The MesserschmittBf 110 C–4/B of I./ZG 1 *Wespen*, Eastern Front, spring 1942.
(MVT via M. Krzyżan)

The Messerschmitt Bf 110 C–4/B of ZG 1 *Wespen* during bomb loading, Eastern Front, winter 1941/42.
(P. Jarrett via B. Ketley)

252 Sqns. RAF claimed more than 10 Junkers. The Bf 110 was claimed by the No. 250 Sqn C.O., S/L Waddy. On May 27, 7./ZG 26 lost two Bf 110s (including that of Uffz. Arno Frickmann). The same day Olt. Johannes Berfleth's machine was downed over Tobruk. During an attack against the airfield on June 1, 1942 Olt. Alfred Wehmeyer, a veteran of the unit (18 kills) was killed, as was Uffz. Karl–Heinz Biner, a gunner. Wehmeyer was flying the Bf 110E–1 (3U+NT). During the battle of Gazala, the Bf 110s supported the ground units in many cases, bombing enemy armoured columns. In June 1942, III./ZG 26 lost a Bf 110 which, damaged by AA artillery, ditched in the sea, where the gunner, Uffz. Kunke drowned. On July 5, 1942, despite a strong cover of III./JG 53 fighters, a Bf 110 was shot down by a *Kittyhawk*. The same day two Bf 110 were claimed by Sgt Stephenson and F/L Westlake of No. 450 RAAF Sqn. On the night of July 7, British artillery was shelling the Fuka airfield of III./ZG 26, but none of the aircraft was damaged. On July 9, 1942 No. 252 Sqn *Beaufighters* attacked Bf 110s flying escort to Ju 52/3m between Crete and Tobruk. Two days later another combat ensued. This time the III./ZG 26 lost two Bf 110s. In the afternoon five Italian Fiat CR.42s attacked by mistake a Bf 110s of III./ZG 26. A recce Bf 110F–2 of 4.(H)/12 crashed at landing, killing the pilot Uffz. Kurt Schwerdt and the gunner. Another machine of the unit shot down that day carried onboard Olt. Josef Hofbauer, the NSDAP education–propaganda officer of the II./JG 27, who was earning this way his flying premium and decorations. Hofbauer reported that when he baled out, a Curtiss was circling round with the pilot waving his hand. On July 28, 1942 Bf 110s of 9./ZG 26 shot down a No. 203 RAF Sqn *Maryland* off Crete, and when attacking a convoy, a No. 459 RAAF Sqn Lockheed *Hudson* was destroyed. An attack of the Special Air Service at Quasaba airfield inflicted damage on two Bf 110s.

On August 20, 1942, the III./ZG 26 had a total of 46 Messerschmitt Bf 110Ds and Es. In August the III./ZG 1 arrived in Africa with Bf 110Fs and Gs. The unit was used mainly for bombing attacks. Soon after arrival in Africa the III./ZG 1 Gruppenkommandeur, Maj. Roland Bohrt, was killed in a combat with British fighters. On November 1942, the III./ZG 26 lost a plane which hit an ... anti–tank mine during a forced landing. In November and December 1942 the Messerschmitts were operating from Tunisian aifields. In the Spring of 1943 the III./ZG 26 was transferred to Sicily. In April also the II./ZG 1 arrived at Sicily, transferred from the Eastern Front.

Another chapter of the Bf 110 fighting career was the participation of 4./ZG 76 Bf 110D–3s in the Iraqi rising in May of 1941. The Sonderkommando 'Junck' flew combat missions in the Mosul area in northern Iraq.

OPERATION 'BARBAROSSA' – 1941

Only two Messerschmitt Bf 110–equipped Gruppen took part in the attack on USSR. These were II./ZG 26 'Horst Wessel' and the new SKG 210 (Schnellkampfgeschwader 210) assigned to the Luftflotte. Also a flight of JG 5 (ex–JG 77) equipped with '110s' flew in Norway. During the first days of the attack the Messerschmitts mainly attacked the Soviet airfields. The 5./ZG 26, operating from Suwałki attacked Szawle and Oszmiana. In a couple of days the Messerschmitts 'grounded' some 650 red–star aircraft. On August 19, 1941 the II./ZG 26 attacked Nizhevo airfield, some 50 km south of Leningrad and destroyed 15 aircraft there, while damaging 30 more. Three Polkarpov I–16 fighters of 47 IAP (No. 47 Fighter Regiment) were shot down in the air combat over the airfield. During the operation 'Beowulf', Bf 110s attacked enemy positions in Saarema and Hiuumaa islands. The Messerschmitts of 13.(Z)/JG 5 'Eismeer' attacked the Vaenga near Murmansk. The SKG 210 operated in the Moscow area and e.g. attacked Tula.

In 1942 more Bf 110 units were sent to the Eastern Front. These were I. and II./ZG 1. At the end of 1942 the I./ZG 1 received the first Bf 110G–2s; a Bf 110G–2 S9+GB, was written off

continued on page 31

The Messerschmitt Bf 110 C–5 of long range reconnaissance unit.
(MVT via M. Krzyżan)

Bf 110 C 'U8+HL' of I/ZG 26 during the Battle of Britain, 1940. Aircraft is finished in dark green (RLM 71), black-green (RLM 70) and sky blue (RLM 65). The white nose and strip around the rear fuselage were elements of quick identification system in this period.

Bf 110 D–1/R1 'M8+SL' of 3./ZG 75, Aalborg, Danmark, August 1940. Aircraft is finished in RLM 70/71/65 camouflage.

Bf 110 C–6 'S9+RH' of 1./*Erprobungsgruppe* 210, Denian, France, July 1940. Aircraft is finished in RLM 70/71/65 camouflage.

Bf 110 D–3 of 4./ZG 76 based in Iraq, 1941. Aircraft is finished in RLM 74/75/79/65 camouflage.

Bf 110 D–3 '3U+KT' of 9./ZG 26, Northern Africa, 1941. Aircraft is finished in RLM 74/75/76 camouflage.

Mal. Jacek Jackowski

Messerschmitt Bf 110 G–2/R3/M1/M5 '2N+BP' of 6./ZG 76, Germany, March 1944. This aircraft was used in Third Reich's territory defence.

Aircraft is finished in grey-green (RLM 74) and grey-violet (RLM 75) on top surfaces and white-blue (RLM 76) on lower surfaces.

Under the fuselage it has installed ventral gun pack with two 20-mm MG 151/20 cannons (M1 modification), under the wings are installed two twin 210 mm W.Gr. 42 rocket launchers (M5 modification). Moreover under the wings the aircraft could carry two auxiliary fuel tanks of 300 l capacity each (*Rüstsatz* 7) and in place of the four MG 17 machine guns it has two MK 108 cannons (*Rüstsatz* 3).

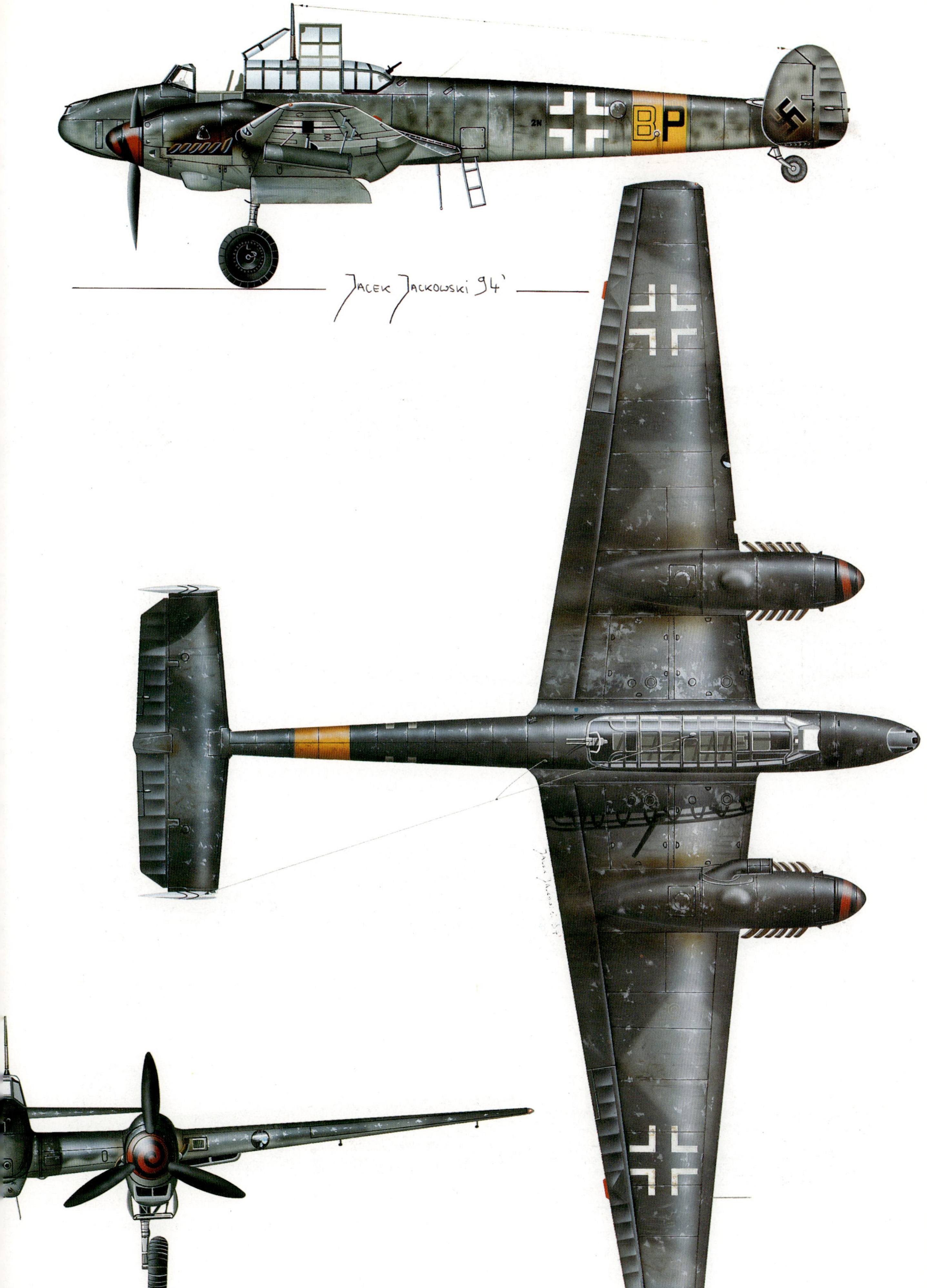

Mal. Jacek Jackowski

Bf 110 G–2/R3 '3U+DT' of 9./ZG 26. Aircraft is finished in RLM 74/75/76 mottle camouflage.

Bf 110 G–4/R3 of I/ZG 1, 'white 12', Bad Lippspringe. Aircraft is finished in RLM 74/75/76 mottle camouflage.

Bf 110 G–2 'G9+XR' of 7./NJG 1, summer 1943. Aircraft is finished in RLM 76 overall. Lower surfaces of the starboard wing, fuselage and tail is finished in RLM 22 (black). This scheme of camouflage was carried by *Beleuchtergruppe*'s aircraft which were used to droping flares above formations of Allied bombers.

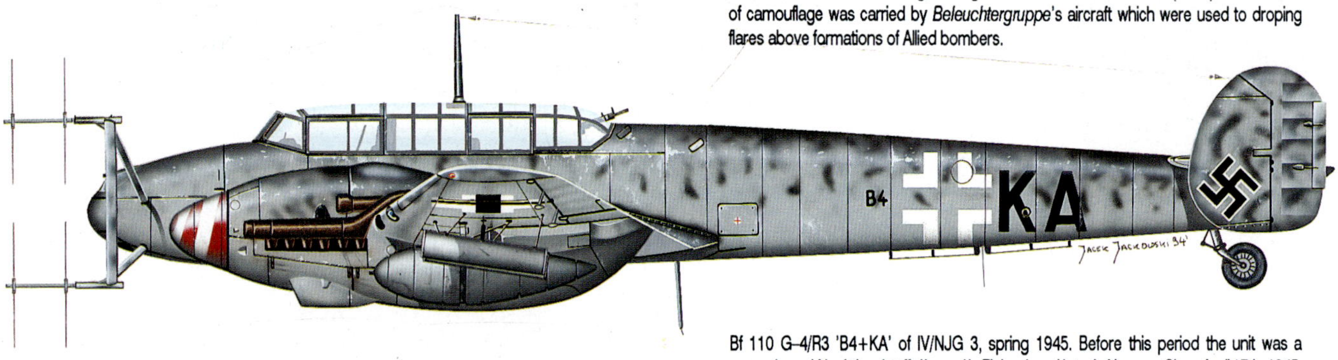

Bf 110 G–4/R3 'B4+KA' of IV/NJG 3, spring 1945. Before this period the unit was a unnumbered *Nachtjagdstaffel* based in Finland, and later in Norway. Since April 17th, 1945, unit had been incorporated in NJG 3 as IVth Gruppe. Aircraft has a FuG 220 *Lichtenstein* SN2b radar. It is finished in RLM 74/75/76 mottle camouflage.

Bf 110 G–4/R3 'D5+BR' of 7./NJG 3, late 1944. Aircraft has a FuG 220 *Lichtenstein* SN2c radar.It is finished in RLM 74/75/76 mottle camouflage.

Mal. Jacek Jackowski

The Messerschmitt Bf 110F on the front airfield, probably Eastern Front.
(MVT via M. Krzyżan)

in a landing accident. Bf 110s were also delivered to dive–bombing units, e.g. two Bf 110s to the Stab of StG 77, or to the recce units – a Bf 110 in NAGr (Nahaufklärungsgruppe – close reconnaissance squadron). A Bf 110G–2 (H8+HM) was destroyed at Leningrad. The Bf 110s were also used against tanks, for example by ZG 1. Later in the war the BF 110 units were withdrawn to the Reich, or re–equipped with other types, such as the Me 410 or Fw 190A/F. The 2.(H)/33 reconnaissance flight was destroyed at Stalingrad. In 1944 the Messerschmitt Bf 110G–2s of 13.(Z)/JG 5, under Hptm. Herbert Treppe, were used against British air cover for the Murmansk convoys.

5./NJG 200 operated from the Odessa airfield. The unit carried out Helle Nachtjagd (free night hunting). Moreover, the night fighters flew day patrols over the Black Sea.

REICH AIR DEFENCE 1941 – 1945

The use of Messerschmitt Bf 110s in day and night fighting Reich air defence units is yet another story. During July 1940 the first night fighting unit, I./NJG 1, was created. At first the Messerschmitt Bf 110Cs were guided from the ground. In 1942 the aircraft with on – board radar stations were introduced.

The night fighting units operated not only over the German territory, but also on the Eastern Front, Romania and Italy. NJG 100 flew on the Eastern Front, the NJ – Finland in Finland, NJG 4 in Romania and IV./NJG 6 in Hungary. The Bf 110s took part in defence against the mass raids of the American aircraft attacking the industrial centres of the Reich in 1943.

The leading German pilots, Werner Streib and Wolfgang Falck, achieved their greatest successes when flying Messershcmitt Bf 110s. The Bf 110s used various bomber interception tactics when operating at night. Usually ground guidance stations were used, although the crews also operated individually above their RAF targets. Olt. Kurt Martinek, a pilot of 1./NJG 4, shot down four RAF aircraft: two *Wellingtons*, a *Stirling* and a *Halifax* between August 25 and September 3 while flying a Bf 110E. The leading Luftwaffe ace, Schnaufer, achieved no less than 121 kills.

The Reichsverteidigung aircraft were armed with additional Mauser MG 151/20 cannon fitted under wings and the fuselage. The Bf 110Gs could also accomodate the 'Schräge Musik', the MG FF/M cannon fitted in the fuselage to shoot upwards.

The Bf 110s were also used to intercept the DH98 *Mosquito* bomber and reconnaissance aircraft. III./NJG 5 under Hptm. Paul Zorner, and based at Mainz–Finhein, specialized in *Mosquito* hunting. On the night of April 21, 1943 at 2:47 Paul Zorner shot down a *Mosquito* while flying a Bf 110G–4, D5+BS. The 9./NJG 5 Staffelkapitän's aircraft was shot down during that combat.

The night–fighting Bf 110s were used by the leading Luftwaffe pilots, e. g. Maj. Helmut Lent

NJG 1 – Air Victories From July 26, 1940 Until The End Of The War

Element	By Day	By Night
Stab NJG 1	6	24
I./NJG 1	–	585
II./NJG 1	119	405
III./NJG 1	7	343
IV./NJG 1	13	816
Total	145	2173

Above: The Messerschmitt Bf 110 D–3 with additional fuel tanks of 900 litres capacity each, probably from training unit that is manifested by painted on the fuselage number only. On the nose and under wings visible designations of 8./ZG 26 for that plane previously belonged.
(MVT via M. Krzyżan)

Below: Destroyed and partly burned Messerschmitt Bf 110 G–2 of unknown unit.
(MVT via M. Krzyżan)

The Messerschmitt Bf 110 C–4 (CT+OB), 'red 10' after forced landing on one of the Hungarian airfields, 1944. Note low visible two narrow yellow strips painted after wing trailing edge, designating belonging to blind pilotage school.

(Bernàd Dènes coll.)

Above: The Messerschmitt Bf 110 Cs of German flying school on the Budapest–Ferihegy airfield in Hungary, summer 1944. In foreground aircraft designated K5+AF, behind: DE+MP and CS+NR.

(Bernàd Dènes)

Below: The Messerschmitt Bf 110 C (K5+AF) 'yellow 24' painted on the nose designated flying school, Ferihegy-Budapest, Hungary. Note white spinner.

(Bernàd Dènes)

(102 kills at night and 8 by day), Maj. Wilhelm Herget (57 kills), Maj. Heinz–Wolfgang Schnaufer (121 kills), Hans–Joachim Jabs (22 day kills and 28 night ones).

On April 9, 1943 the crew of the Messerschmitt Bf 110G–4 W. Nr 4811 (G9+CX), Uffz. Heinz Kuhn and Lt. Köstler shot down an Avro Lancaster III (ED618) of No. 101 Sqn RAF. The fire of the gunners damaged the Messerschmitt, which caught fire and crashed; Lt. Köstler was killed while Uffz. Kuhn baled out.

On April 20, 1944 the Bf 110G–4/B2/R8 of Olt. Wilhelm Johnen landed in error at Zürich – Dübendorf. The aircraft was damaged when attacking British *Lancasters* during a raid against the Dornier factory at Friedrichshafen. The machine was interned in Switzerland; to prevent disclosure of the latest FuG 220 radar design, the Germans agreed to deliver 12 brand new Messerschmitt Bf 109G–6 fighters in exchange for the destruction of the Bf 110 in the presence of German officials.

EXPORTS

The Messerschmitt Bf 110 was not as widely exported as the Bf 109 or the He 111.

In 1942 the Italian Regia Aeronautica started forming new night fighting units. Apart from existing night fighter squadrons equipped with Fiat CR.42 aircraft, the new units were to use multi–engined fighters. The Regia Aeronautica Chief of Staff, Gen. Rene Fougnier, placed a contract in Germany for 24 Messerschmitt Bf 110s equipped with FuG 202 *Lichtenstein* radar. Between June and August 1942 three Messerschmitt Bf 110Cs arrived in Italy, where they received serials MM964, 1358 and 1804. Two of these were assigned to 235 Flight of 60 Fighter Squadron, while the remaining one (MM964) was sent to 41 Bomber Regiment which was to receive Fiat BR.20M aircraft fitted with the FuG 202 radar. The Messerschmitts flew from Lonate Pozzolo airfield. Because of the reduction in multi–engined aircraft production (priority being given to Bf 109s and Fw 190s), the Germans cancelled the Italian order for Bf 110s.

After the surrender of Italy some Italian pilots again met the Bf 110; No. 170 Flight of the air force of the so called Social Republic of Italy received from its German allies two new Messerschmitt Bf 110Gs, but the unit flew no operational missions.

At the end of 1944 the Germans delivered several Bf 110F–4s and six Bf 110G–4s to the 101 night fighting flight of the Hungarian Air Force. Flying from Csorne airfield the Hungarian night fighters fought Soviet night bombers over the front. During February the unit was transferred to Wiener Neustadt; the unit fought on until late March 1945.

Rare photograph of a Rumanian Bf 110. The aircraft after crash – landing.

(M. J. Murawski)

The row of Messerschmitt Bf 110 G–4/R3/B2/M2s of 9./NJG 1 with FuG 220 *Lichtenstein* SN2d radar with slanted dipoles. Note that camouflage of one of planes is different. Germany, 2nd of April 1945.

Several Messerschmitt Bf 110s were also delivered to the Romanian air force; two or three Bf 110s of various versions were also used by the Romanians for training.

MESSERSCHMITT BF 110s IN THE USSR

According to some Soviet sources, the USSR purchased several (3 to 5) Bf 110C and D aircraft in Germany in 1940; these were tested at Soviet test centres. During trials a speed of 525 km/h was achieved at 4,600 m altitude, and 442 km/h at ground level; time to 5,000 m was 8.4 minutes. A Bf 110 could make a full turn (360o) in 30 seconds. During 1940 a Messerschmitt Bf 110 was used for testing of the MP–6, 23 mm cannon designed by Y. Taubin and M. N. Babukhin.

In September 1941 a captured Bf 110C was tested in the 158 Fighter Regiment (IAP) od the Air Defence (PVO). The aircraft was also thoroughly tested in the air.

In 1942 a captured Bf 110D was used in the 750 Long Range Regiment (APDD) for recce missions deep beyond the front line. The Messerschmitt was most often flown by P/O Viacheslav G. Opalev and P/O Evgeniy Okorov. During one mission the Soviet Bf 110D shot down a German Ju 88A–4. The aircraft was shot down in error by Soviet AA fire at Tula.

One Bf 110 was displayed in Moscow in 1944 during an exhibition of captured materiel. A Bf 110 fuselage was exhibited in the Soviet Aviation Museum until mid–fifties. In 1993 relatively undamaged Bf 110F fuselage was found in Karelia. This aircraft and another one found in Northern Russia were sold to New Zealand.

Rank equivalents:

German	English	Polish
Gefreiter (Gefr.)	Leading Aircraftsman (LAC)	kapral (kpr.)
Ober Gefreiter (OGfr.)	Corporal (Cpl.)	starszy kapral (st.kpr.)
Unteroffizier (Uffz.)	Lance Corporal (LC)	plutonowy (plut.)
Feldwebel (Fw.)	Sergeant (Sgt)	sierżant (sierż.)
Oberfeldwebel (Ofw.)	Chief Sergeant (CSgt)	starszy sierżant (st. sierż.)
	Cadet Officer (C/O)	podchorąży (pchor.)
	Warrant Officer (W/O)	chorąży (chor.)
Leutnant (Lt.)	Pilot Officer (P/O)	podporucznik (ppor.)
Oberleutnant (Olt.)	Flying Officer (F/O)	porucznik (por.)
Hauptmann (Hptm.)	Flight Lieutenant (F/L)	kapitan (kpt.)
Major (Maj.)	Squadron Leader (S/L)	major (mjr)
Oberstleutnant (Otl.)	Wing Commander (W/Cdr)	podpułkownik (ppłk)
Oberst	Group Captain (G/C)	pułkownik (płk)

The Messerschmitt Bf 110 G–4b/R3/B2/M2 (W.Nr. 73037) captured in 1945, painted with British national insignia. Aircraft was equipped with FuG 220 *Lichtenstein* SN2b and FuG 212 *Lichtenstein* C1 radars and underfuselage ETC 500/IXb bomb racks.

(MVT via M. Krzyżan)

Messerschmitt Bf 110V1

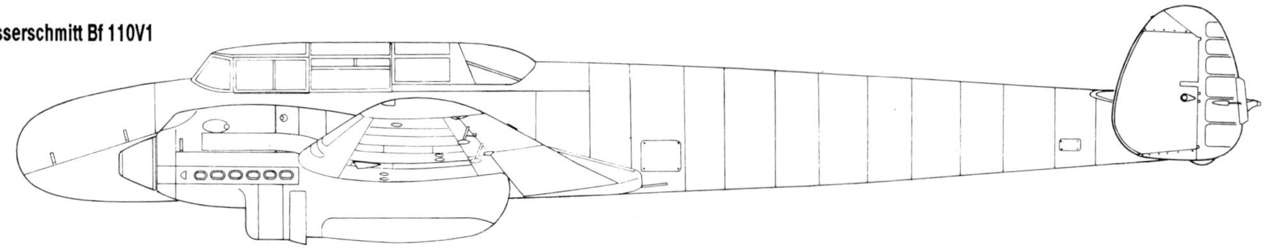

Messerschmitt Bf 110 A–0

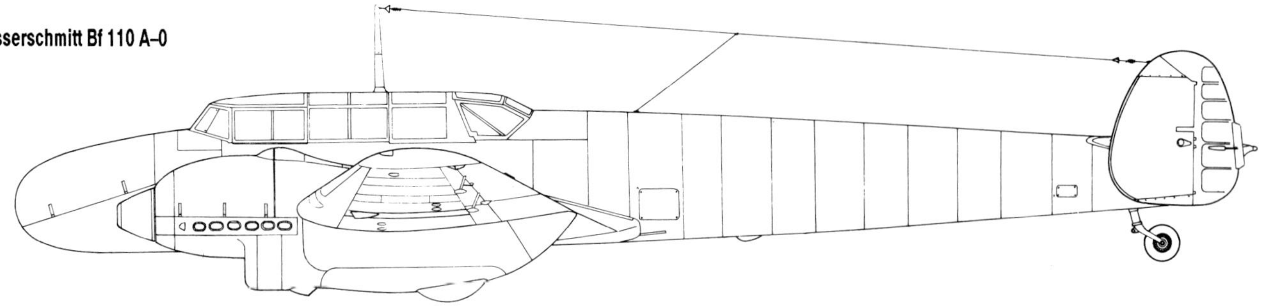

Messerschmitt Bf 110 B–1

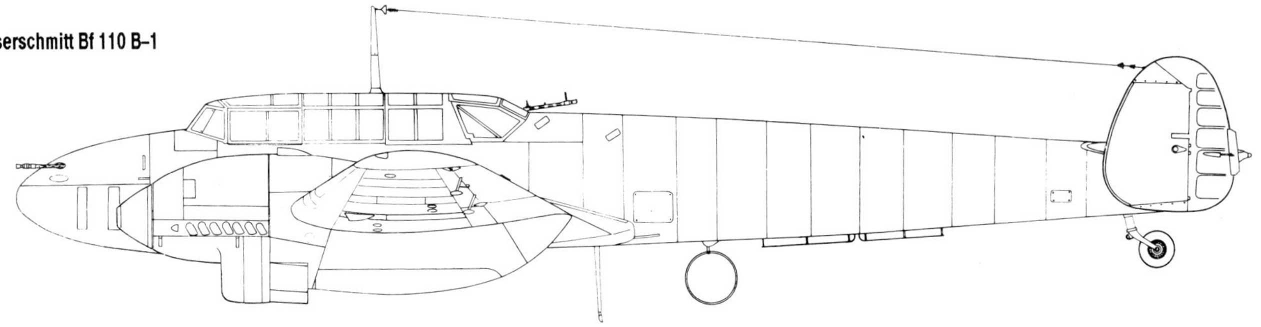

Messerschmitt Bf 110 C–6
the Mk101 cannon under the fuselage

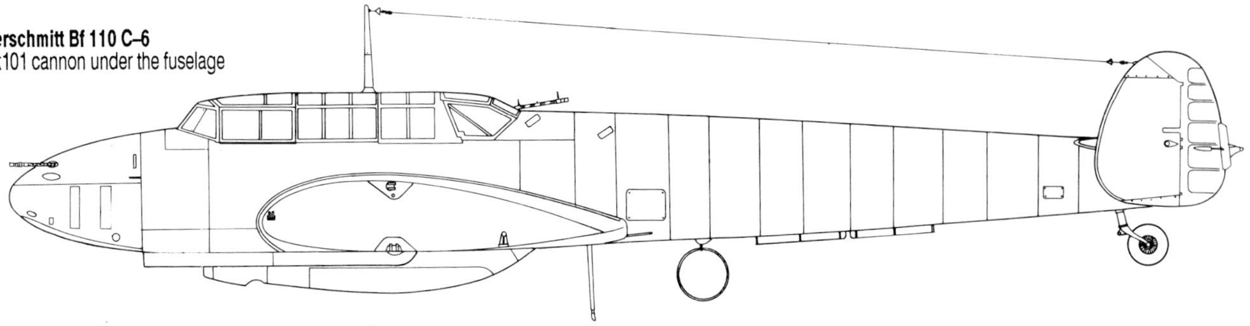

Messerschmitt Bf 110 D–1/R1
the metal fuel tank under the fuselage

Messerschmitt Bf 110 D–1/R1
the wooden fuel tank under the fuselage

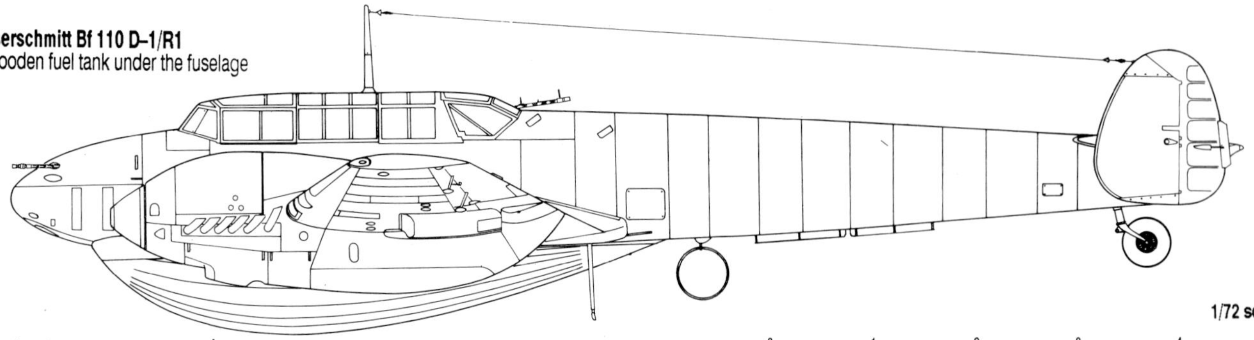

1/72 scale

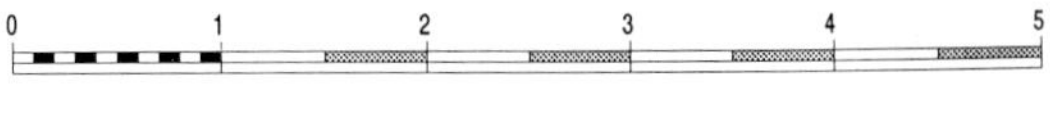

Research and traced by **Krzysztof M. Żurek**

Messerschmitt Bf 110 D–2
underfuselage ETC 500 bomb rack and
additional underwing 300l fuel tanks

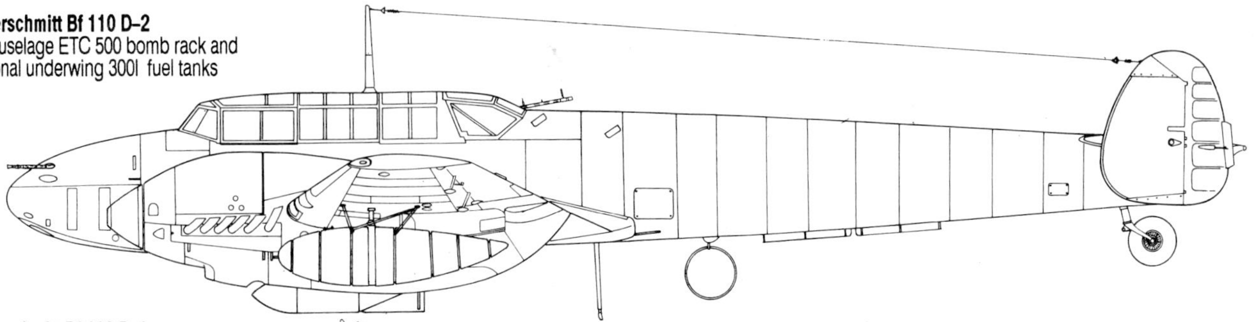

Messerschmitt Bf 110 D–3
extended rear fuselage, additional
underwing 900 l fuel tanks

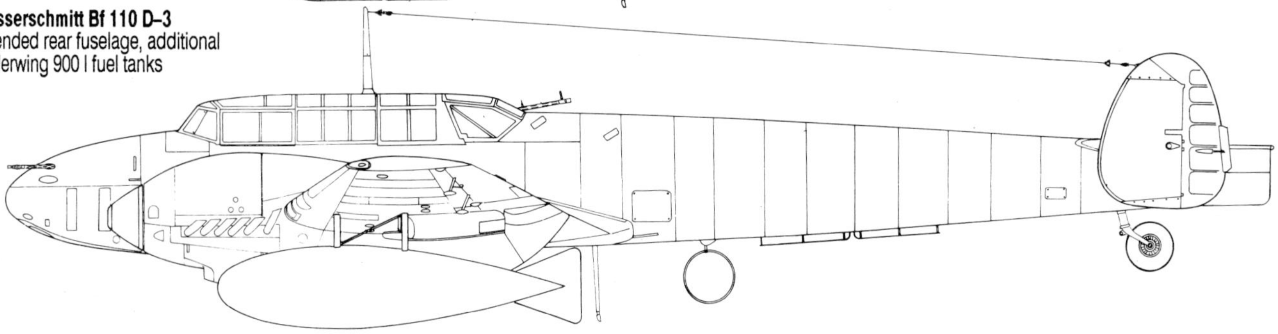

Messerschmitt Bf 110 E–2
D–3 modification with underwing
ETC 50 bomb racks with 50 kg bombs

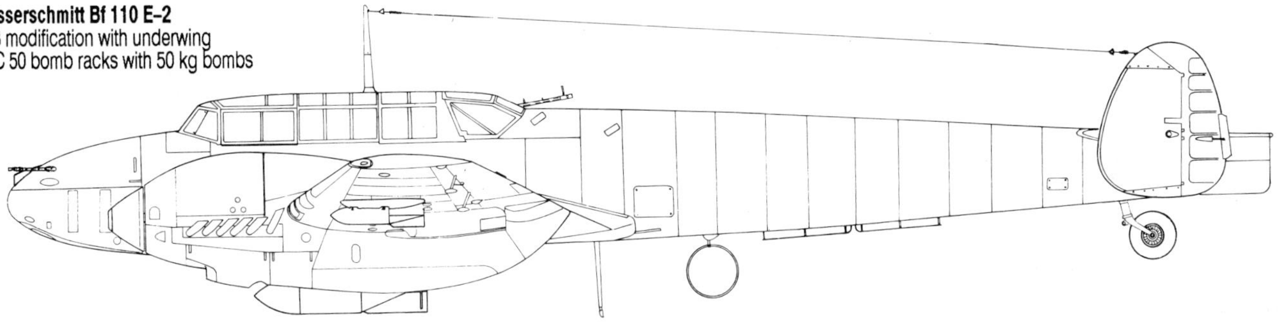

Messerschmitt Bf 110 F–2
W.Gr 21 rocket launchers under wings

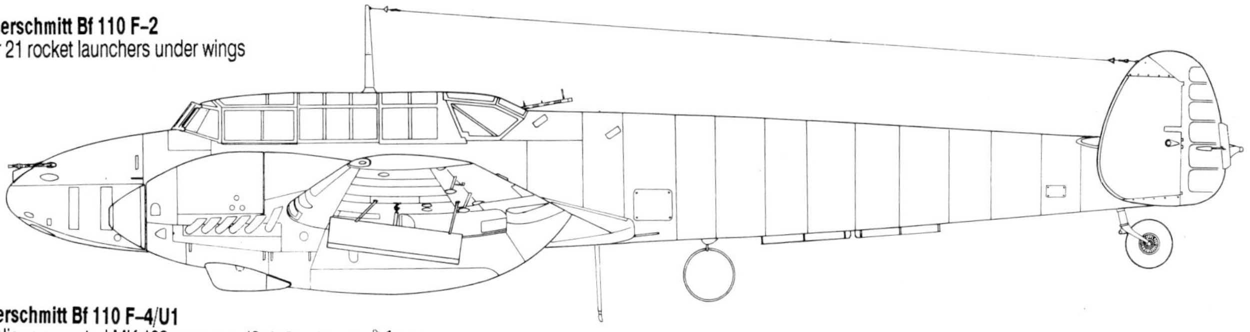

Messerschmitt Bf 110 F–4/U1
two oblique mounted MK 108 cannons *(Schräge Musik)*
in the cockpit

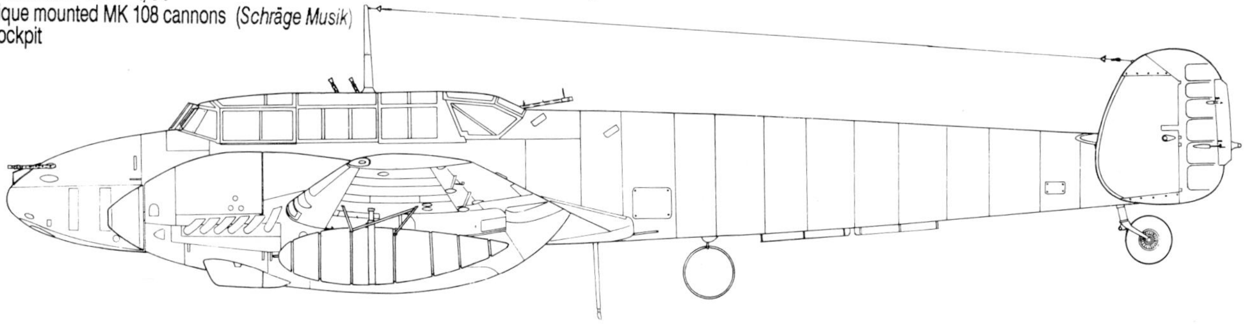

Messerschmitt Bf 110 F–4
underfuselage pod with two MK 103 cannons

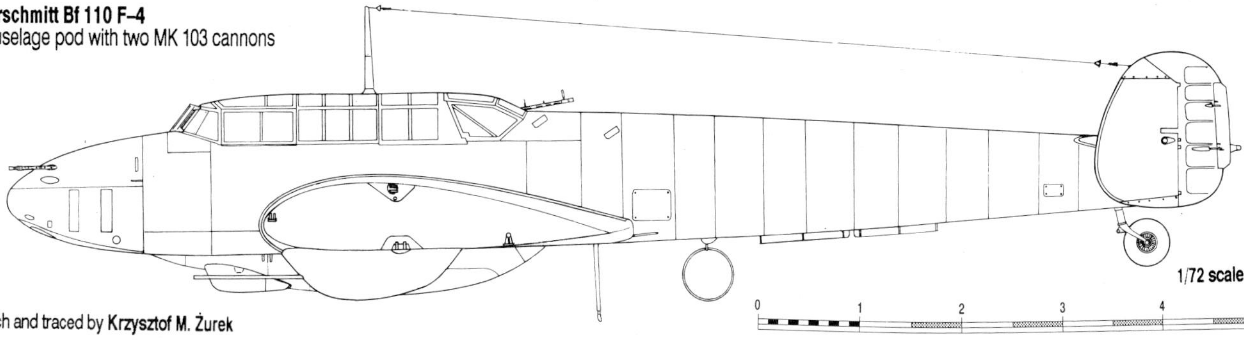

1/72 scale

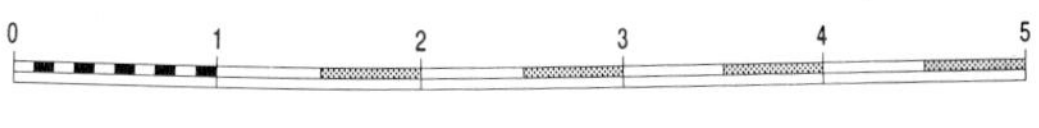

Research and traced by **Krzysztof M. Żurek**

Messerschmitt Bf 110 C–1

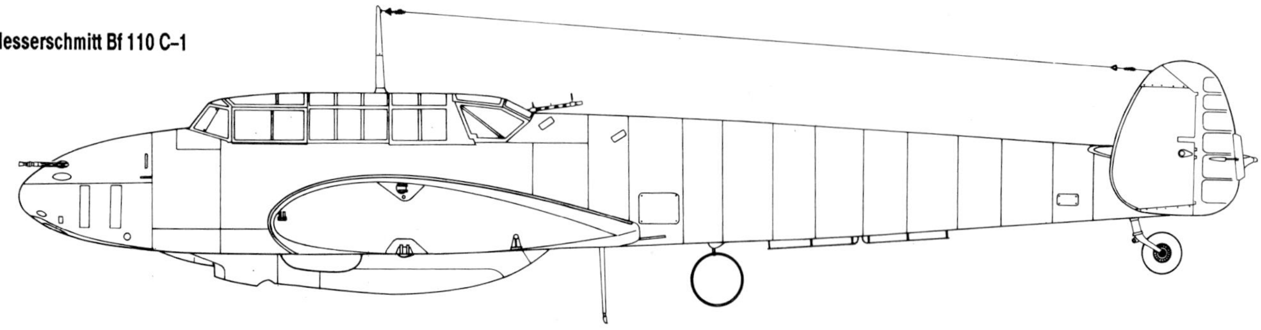

Messerschmitt Bf 110 C–4
front view

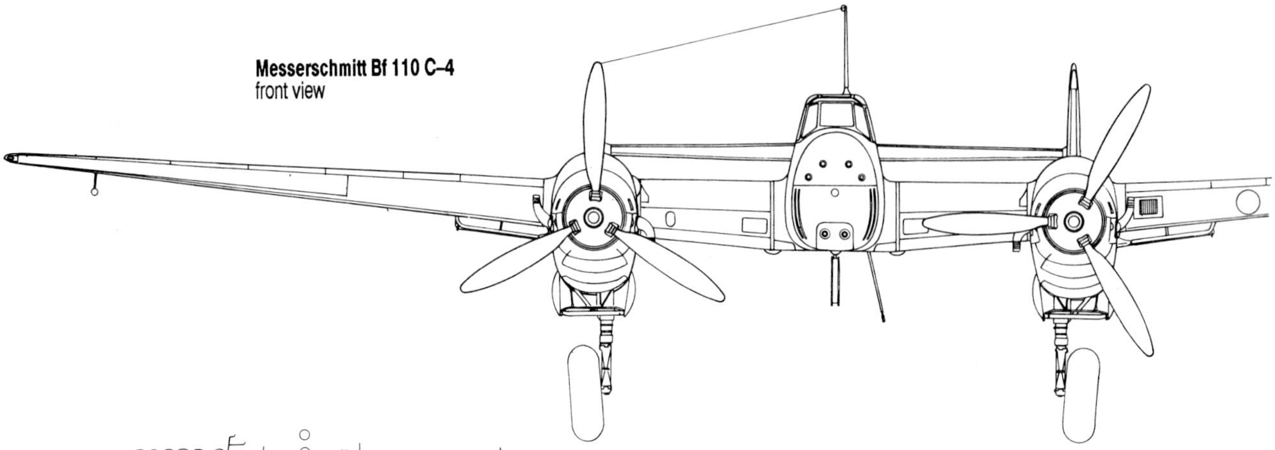

Messerschmitt Bf 110 C–4
bottom view

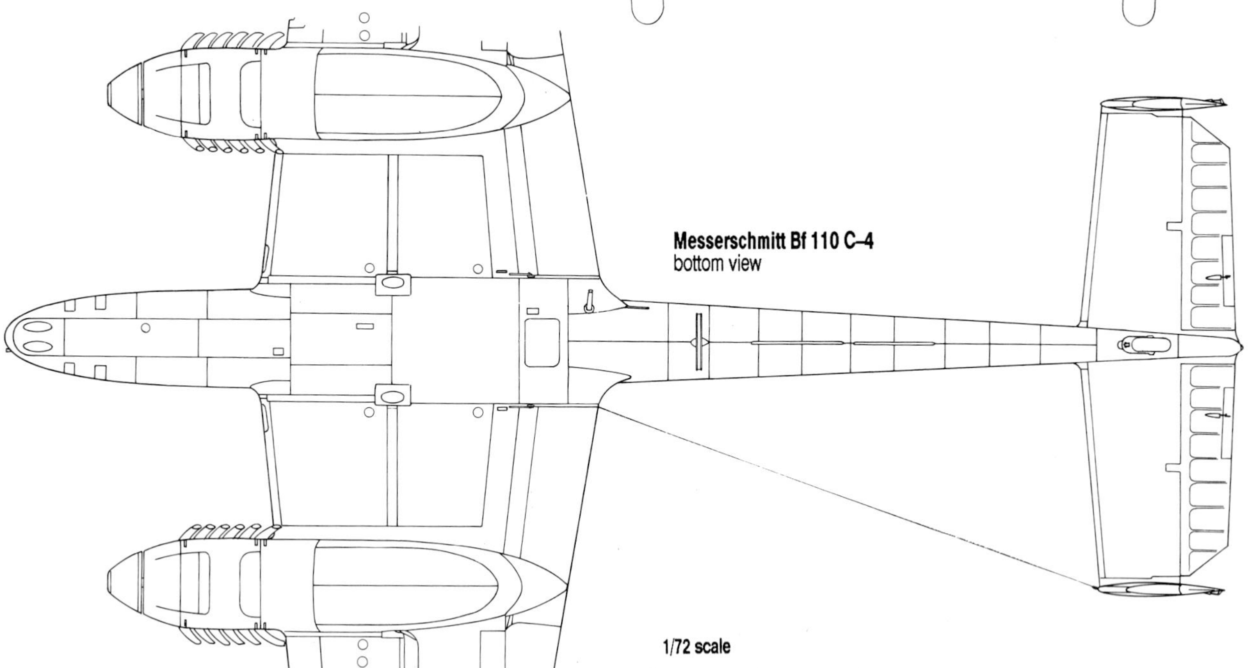

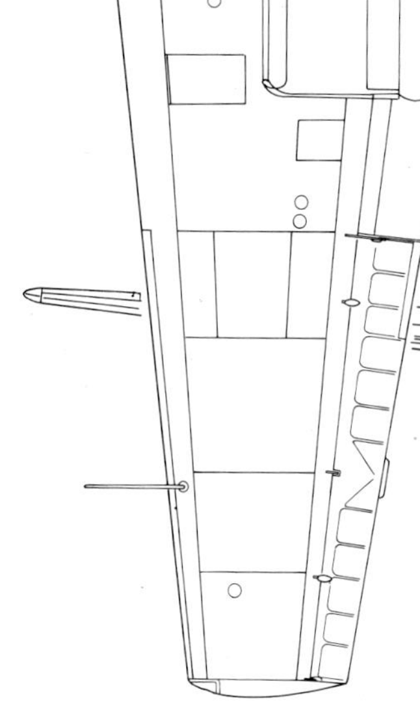

1/72 scale

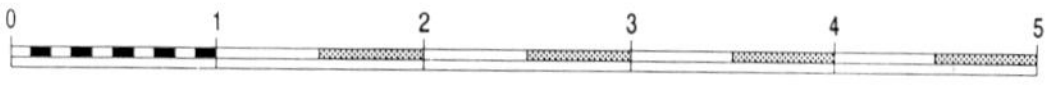

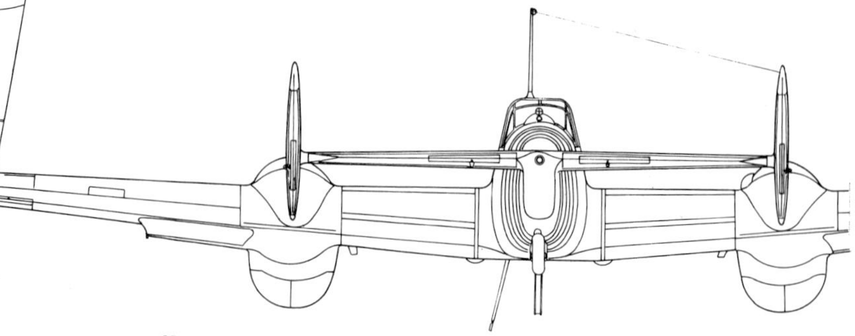

Messerschmitt Bf 110 C–4
rear view

Research and traced by **Witold Hazuka**

Messerschmitt Bf 110 C–4
port side view

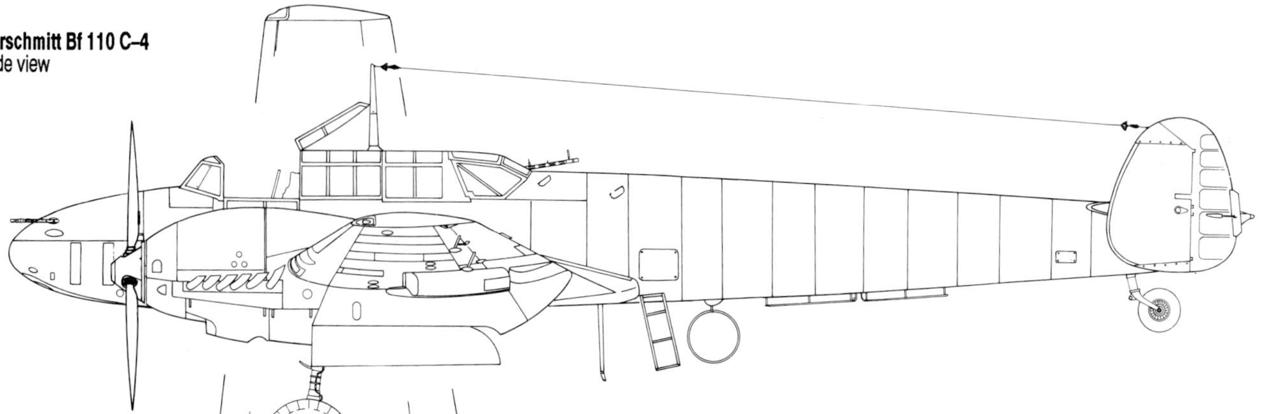

1/72 scale

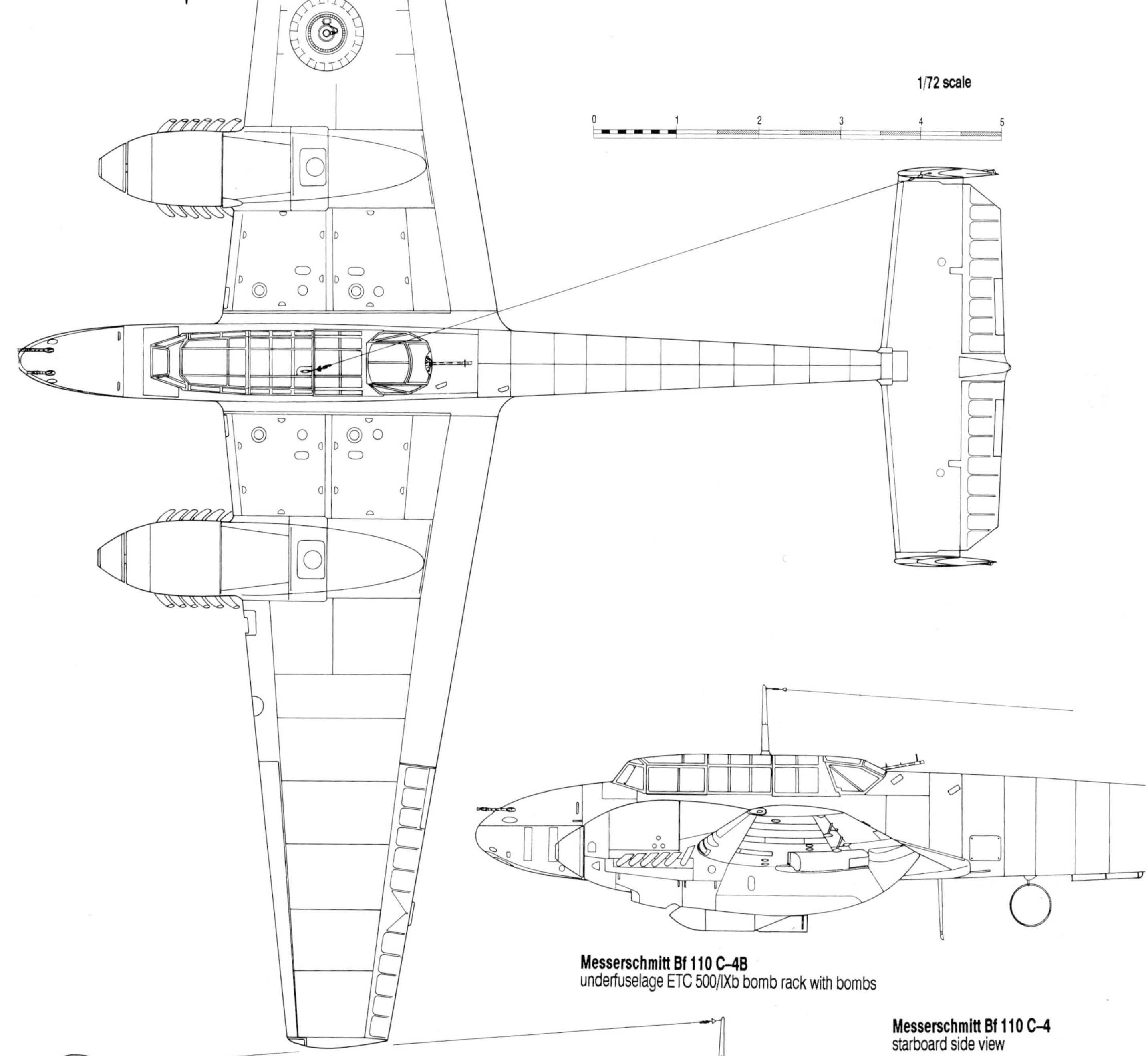

Messerschmitt Bf 110 C–4B
underfuselage ETC 500/IXb bomb rack with bombs

Messerschmitt Bf 110 C–4
starboard side view

Research and traced by **Krzysztof M. Żurek**

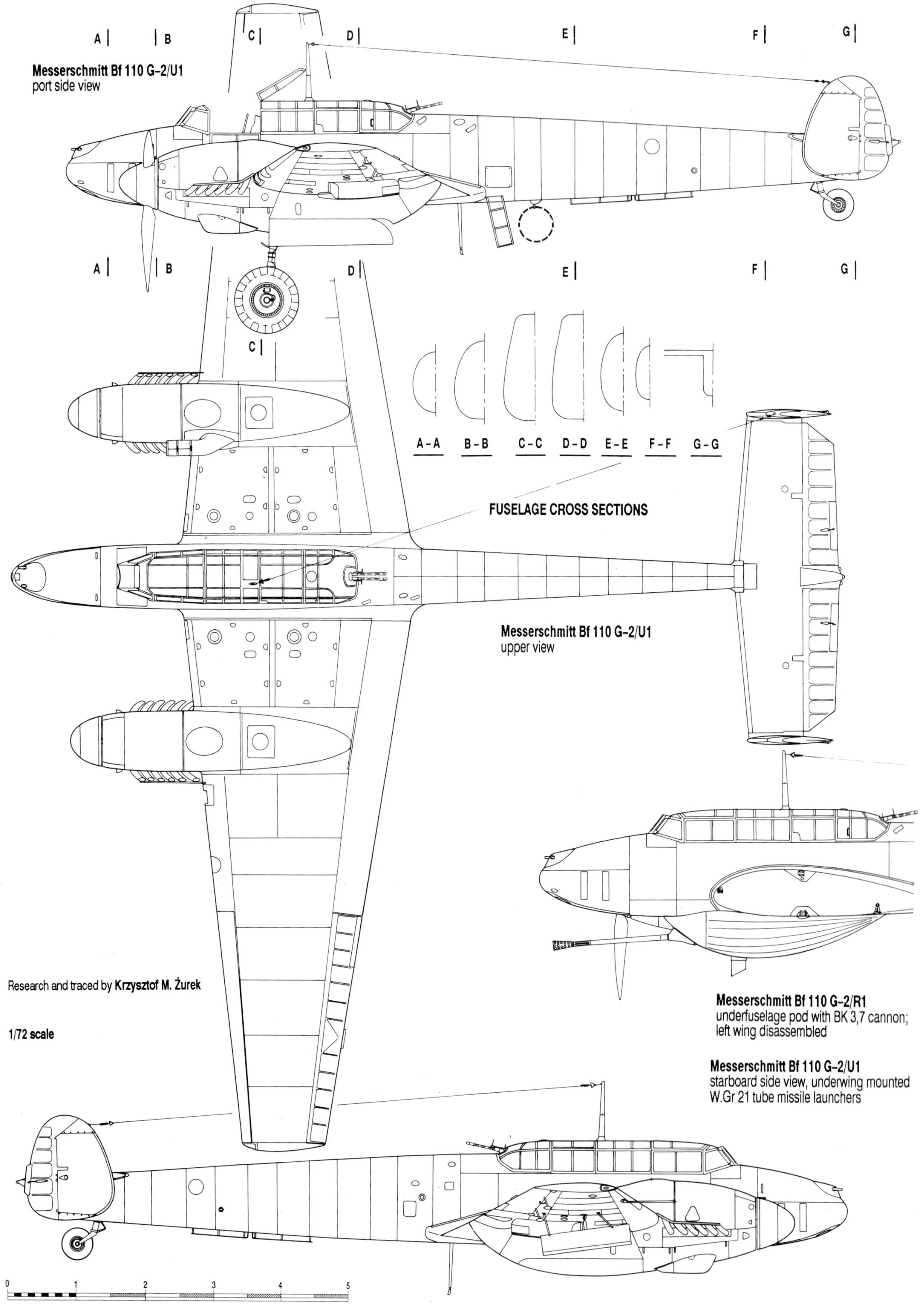

Messerschmitt Bf 110 G-2/U1
port side view

FUSELAGE CROSS SECTIONS

Messerschmitt Bf 110 G-2/U1
upper view

Messerschmitt Bf 110 G-2/R1
underfuselage pod with BK 3,7 cannon; left wing disassembled

Messerschmitt Bf 110 G-2/U1
starboard side view, underwing mounted W.Gr 21 tube missile launchers

Research and traced by **Krzysztof M. Żurek**

1/72 scale

Messerschmitt Bf 110 G-2/U1
front view

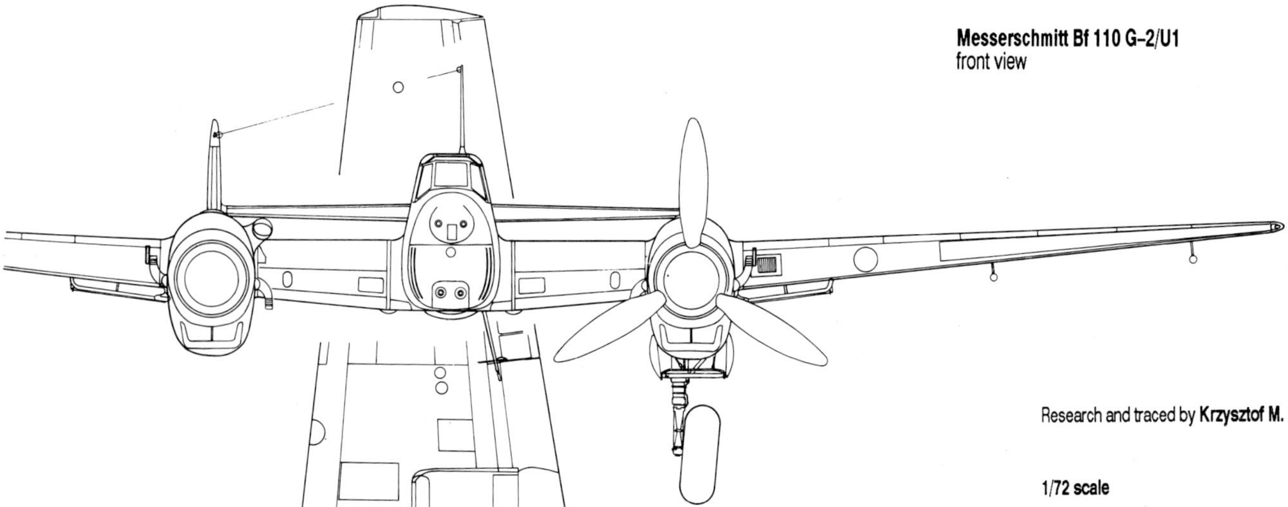

Research and traced by **Krzysztof M. Żurek**

1/72 scale

Messerschmitt Bf 110 G-2/U1
bottom view

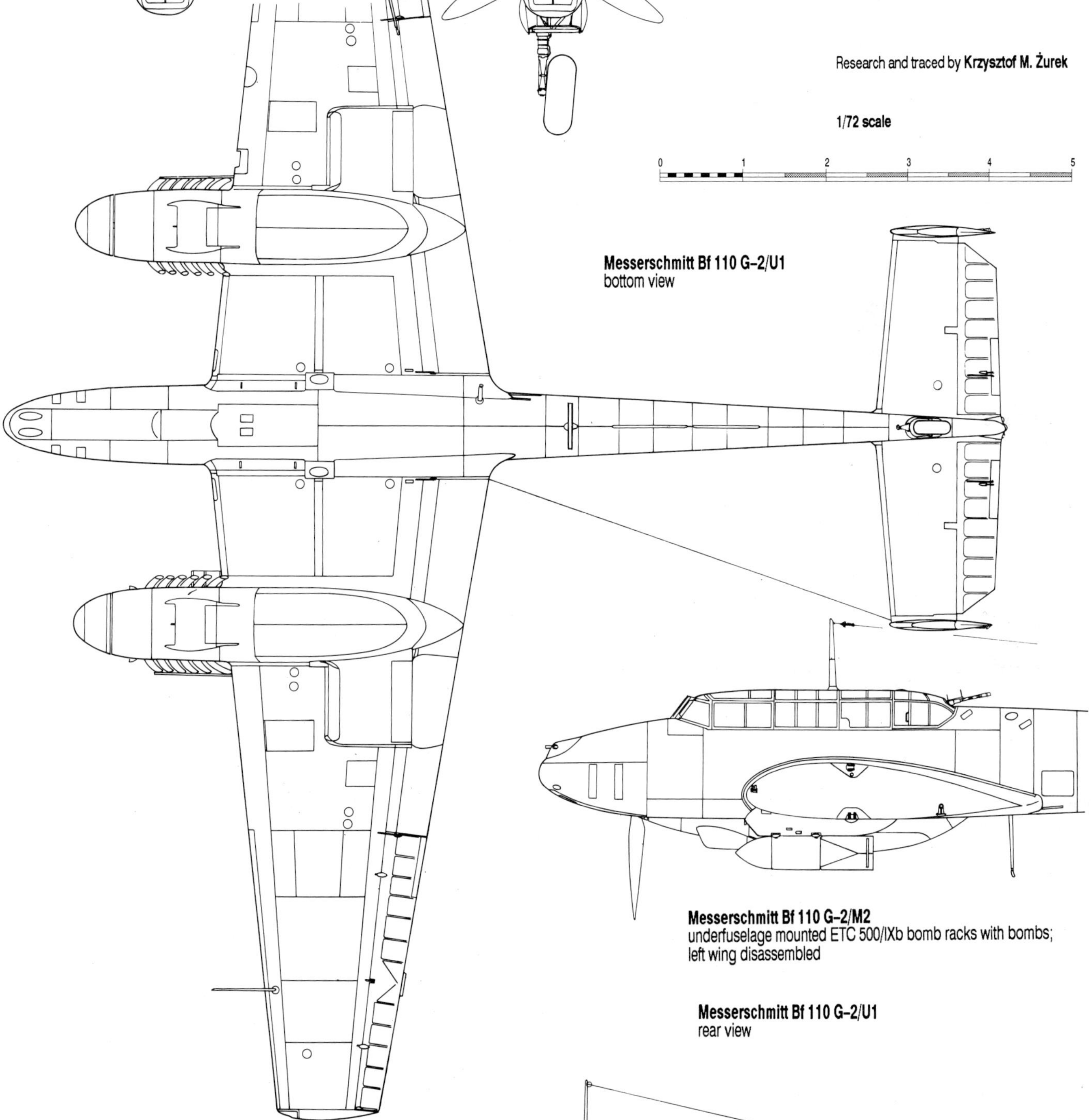

Messerschmitt Bf 110 G-2/M2
underfuselage mounted ETC 500/IXb bomb racks with bombs;
left wing disassembled

Messerschmitt Bf 110 G-2/U1
rear view

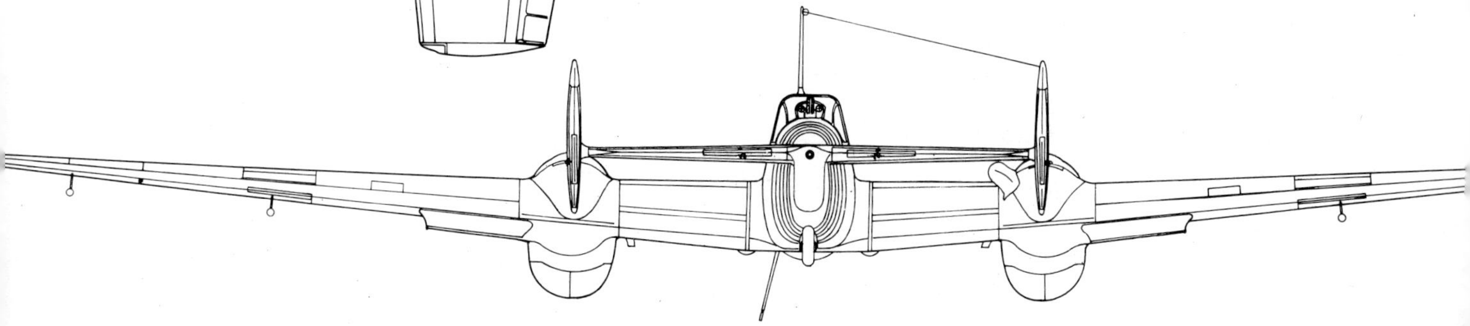

Messerschmitt Bf 110 G–4
with FuG 202 *Lichtenstein* B/C radaraerial

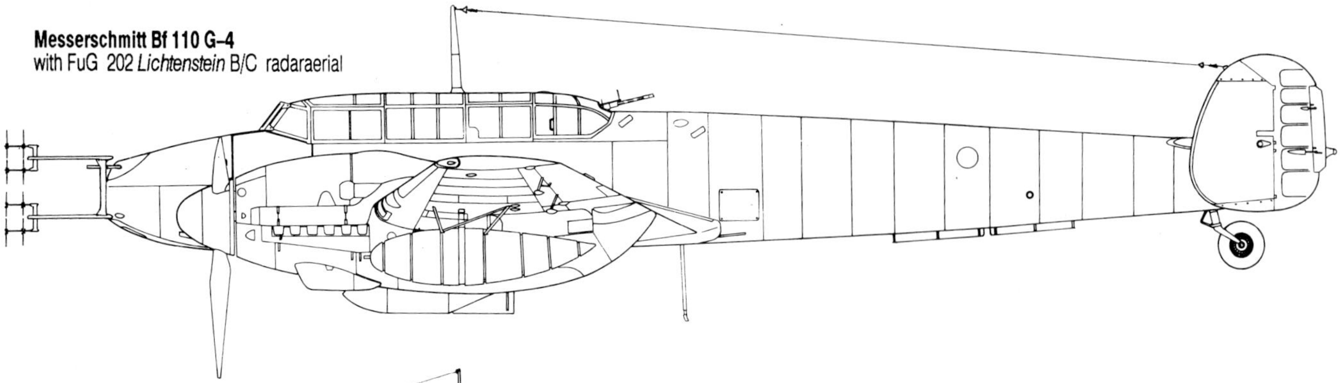

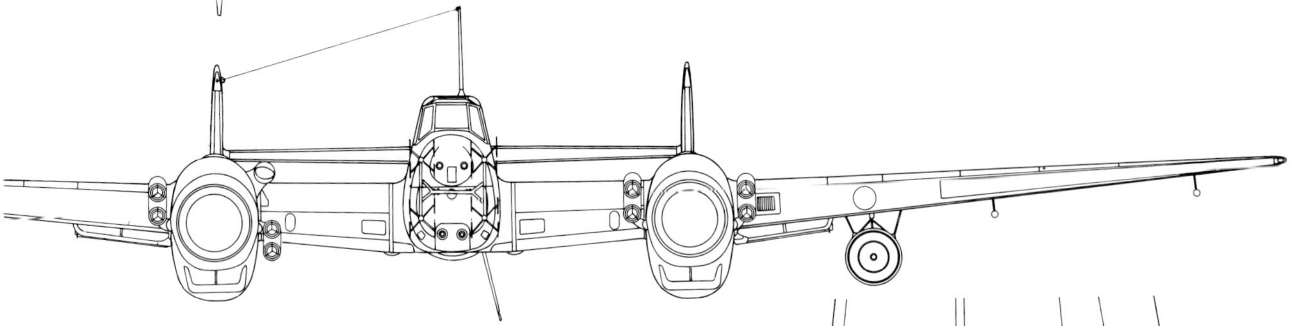

Messerschmitt Bf 110 G–4
with FuG 202 *Lichtenstein* B/C radar aerial
– front and upper views

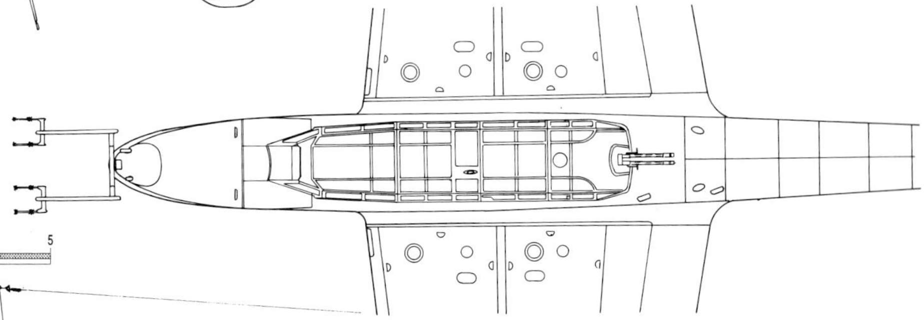

1/72 scale

0 1 2 3 4 5

Messerschmitt Bf 110 G–4
with FuG 212 C1 and FuG 221 *Rosendaal-Halbe* radar aerials

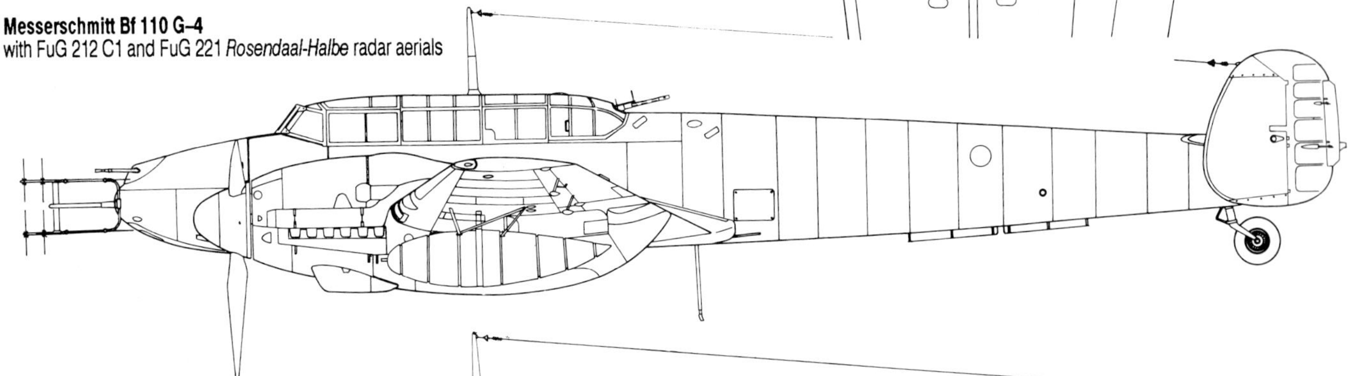

Messerschmitt Bf 110 G–4
with FuG 212 *Lichtenstein* C1 radar aerial

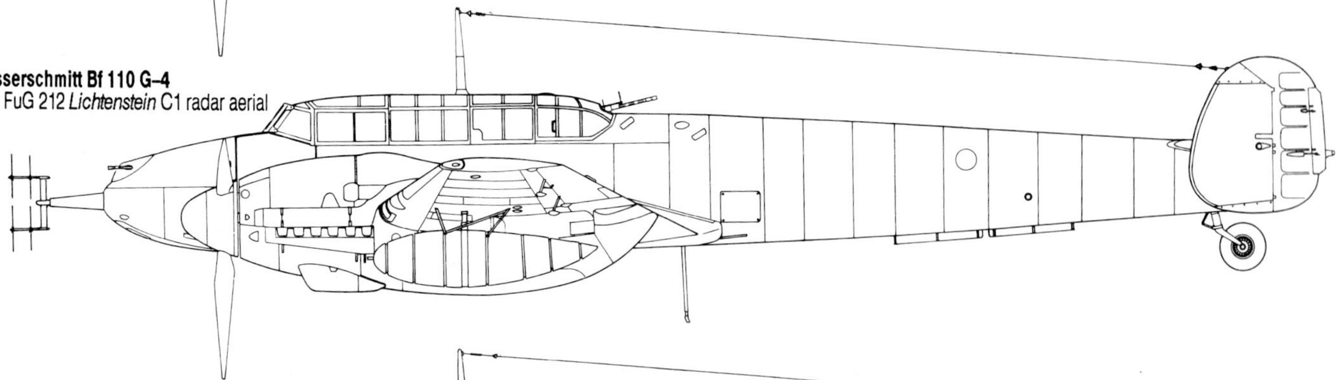

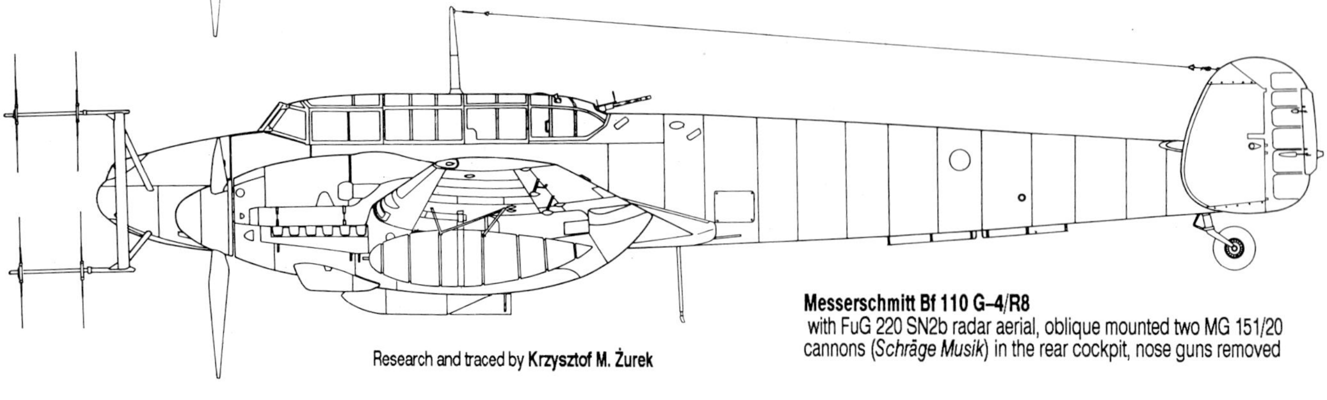

Messerschmitt Bf 110 G–4/R8
with FuG 220 SN2b radar aerial, oblique mounted two MG 151/20 cannons (*Schräge Musik*) in the rear cockpit, nose guns removed

Research and traced by **Krzysztof M. Żurek**

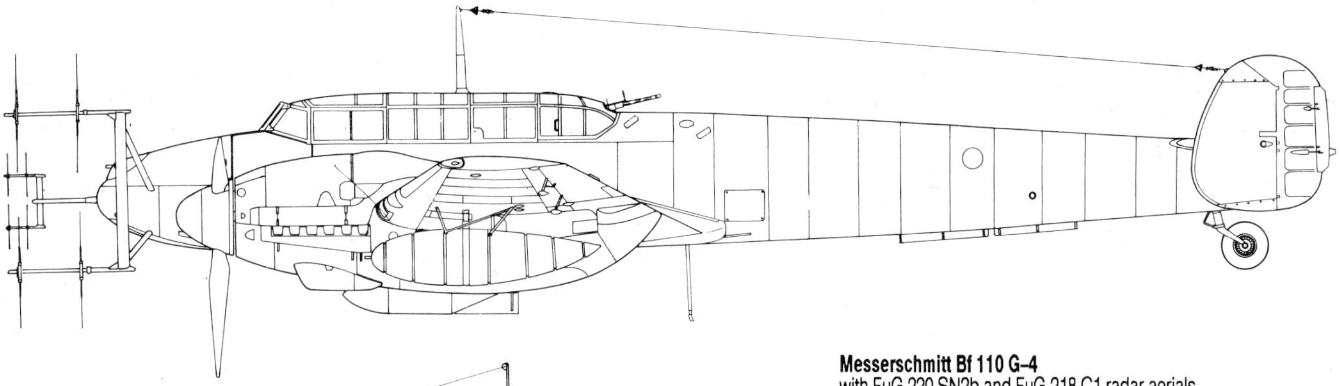

Messerschmitt Bf 110 G-4
with FuG 220 SN2b and FuG 218 C1 radar aerials,
R3/M2/R7 modifications (MK 108 cannons, underfuselage
ETC 500 bomb rack and underwing fuel tanks 300 l capacity each)

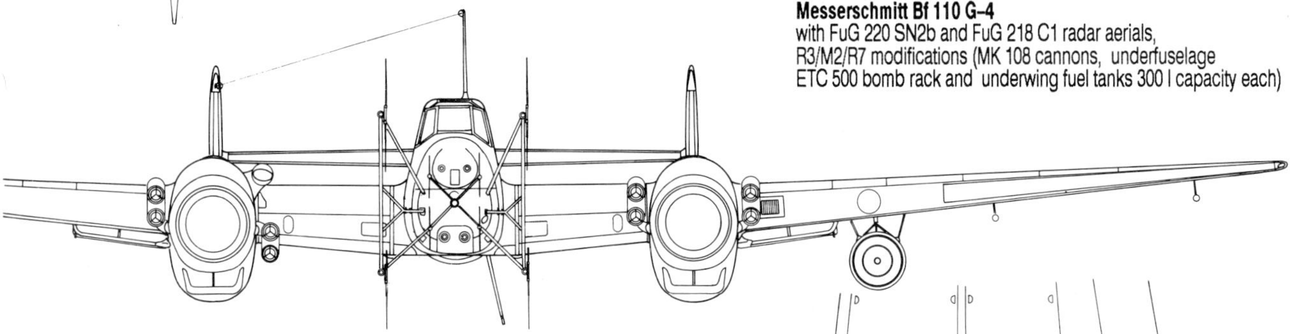

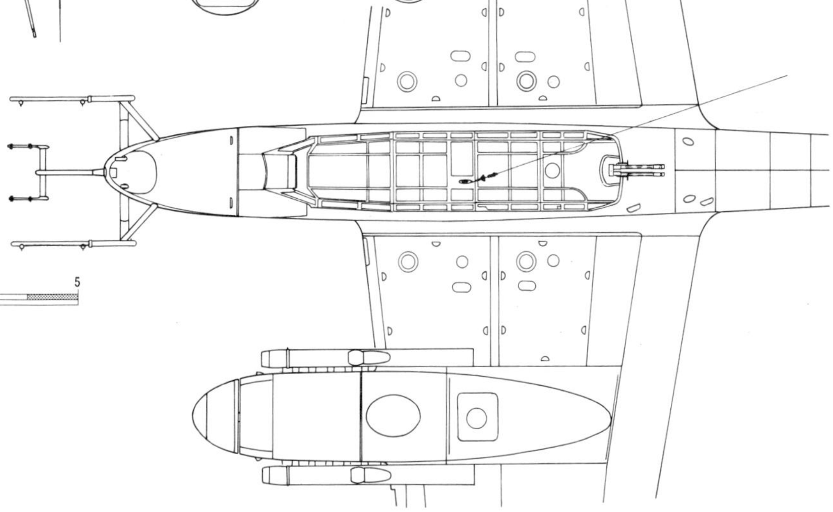

Messerschmitt Bf 110 G-4
with FuG 220 SN2b and FuG 218 C1 radar aerials,
R3/M2/R7 modifications (MK 108 cannons, underfuselage ETC 500
bomb rack and underwings fuel tanks 300 l capacity each)
– front and upper views

1/72 scale

Messerschmitt Bf 110 G-4
with FuG 220 SN2b radar aerials, R3/M2/R7 modifications

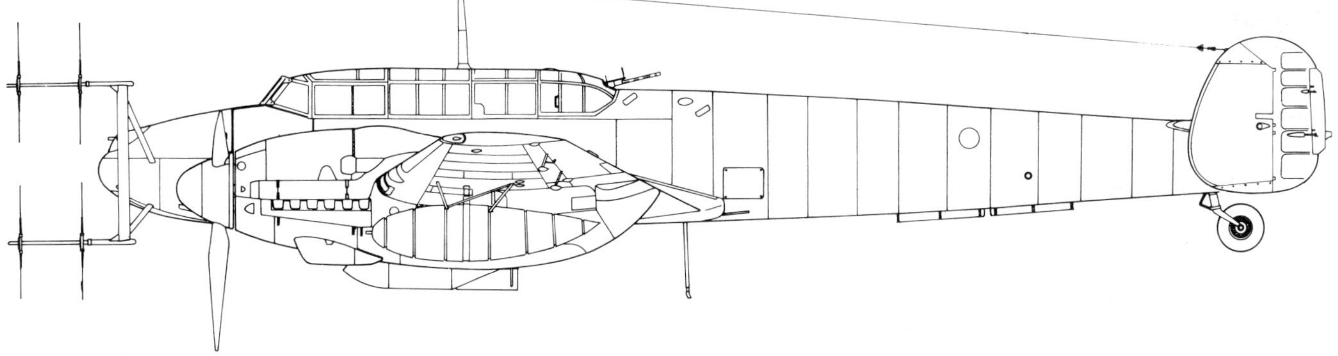

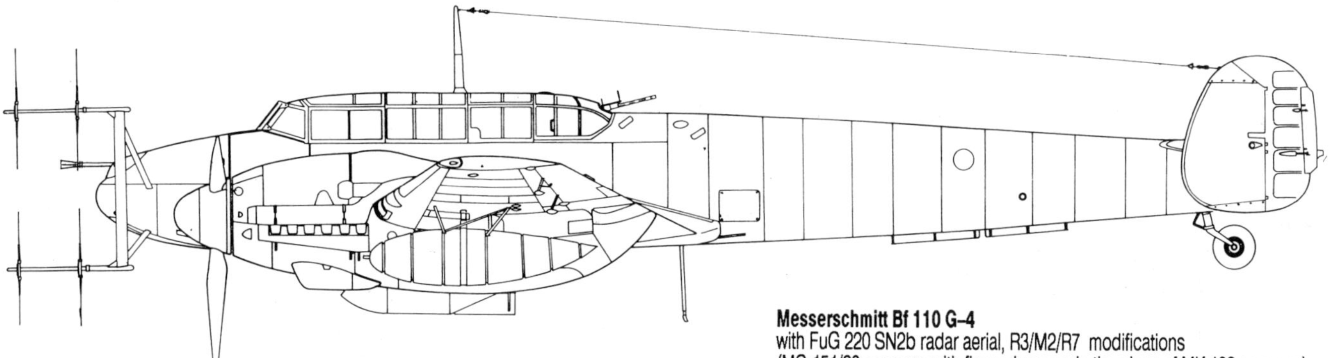

Messerschmitt Bf 110 G-4
with FuG 220 SN2b radar aerial, R3/M2/R7 modifications
(MG 151/20 cannons with flame dampers in the place of MK 108 cannons)

Research and traced by **Krzysztof M. Żurek**

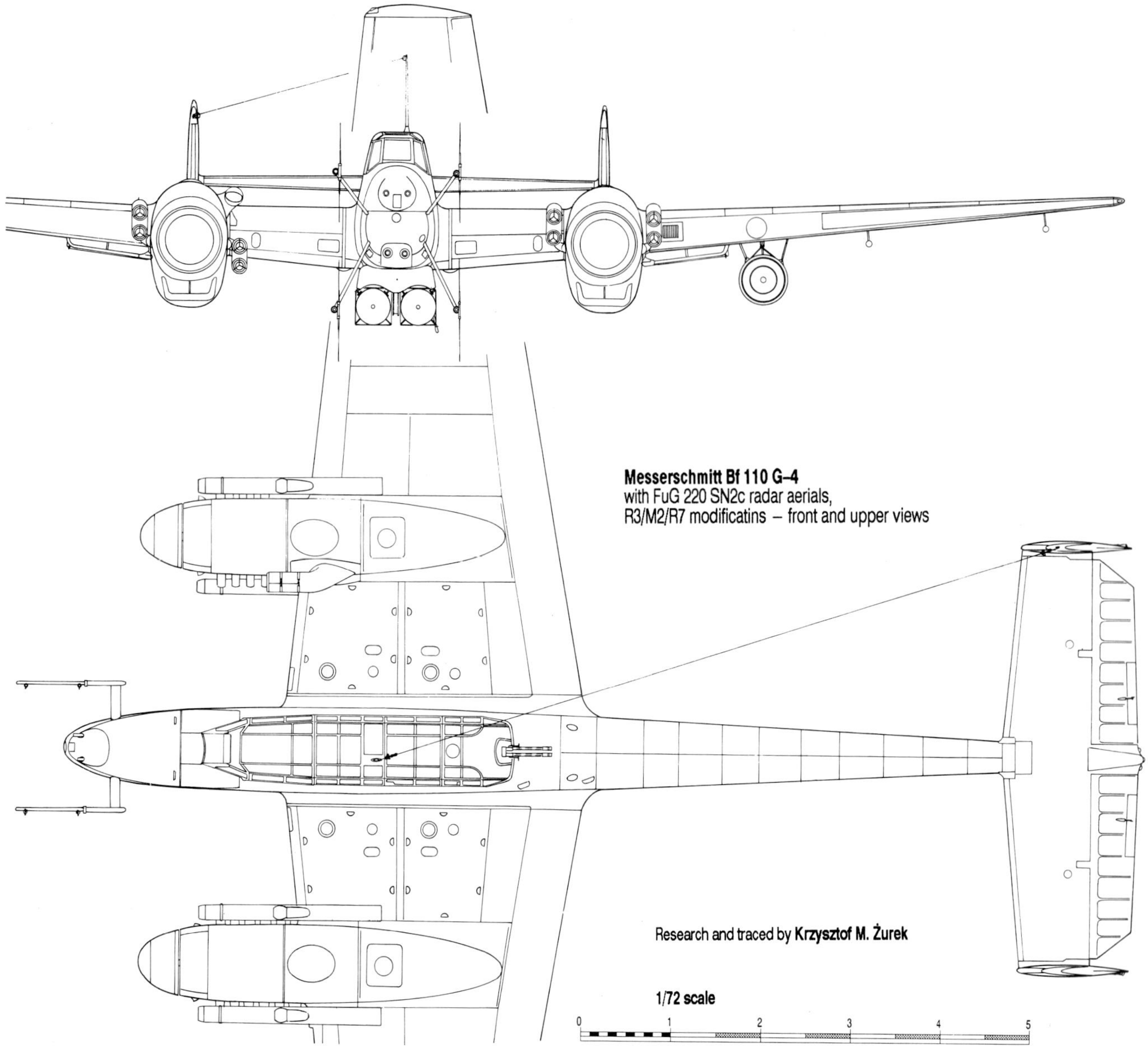

Messerschmitt Bf 110 G–4
with FuG 220 SN2c radar aerials,
R3/M2/R7 modificatins – front and upper views

Research and traced by **Krzysztof M. Żurek**

1/72 scale

Messerschmitt Bf 110 G–4
with FuG 220 SN2b radar, R3/M2/R7 modifications

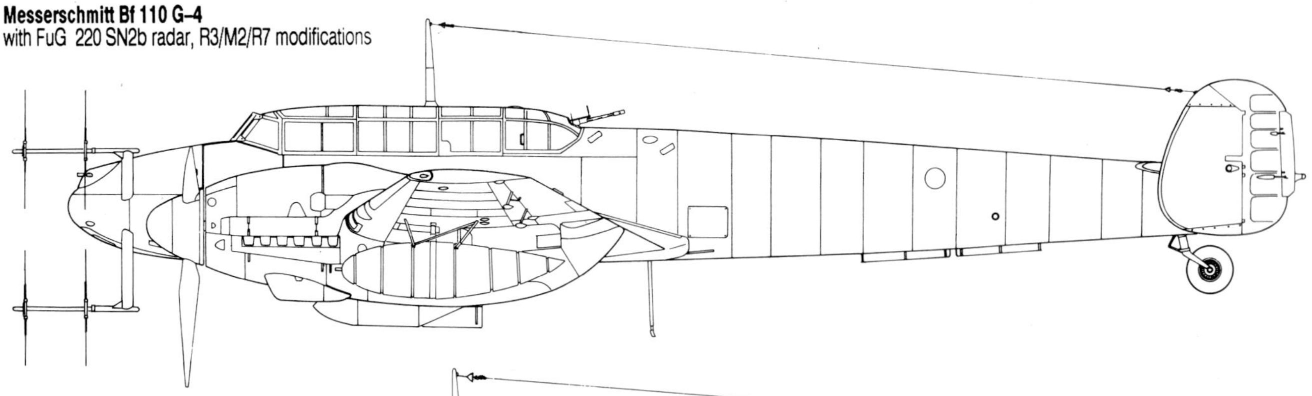

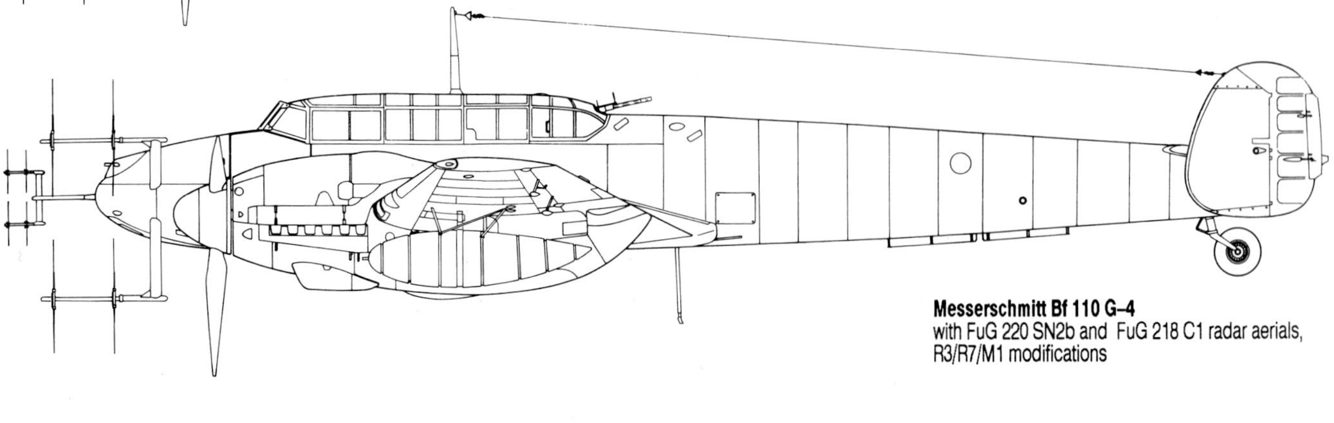

Messerschmitt Bf 110 G–4
with FuG 220 SN2b and FuG 218 C1 radar aerials,
R3/R7/M1 modifications

CAMOUFLAGE AND MARKINGS

The pair of Bf 110 C–3 planes of I./ZG 2 in 1940. Uppermost aircraft (3M+GK) painted in standard RLM 71/70/65 camouflage. Other Bf 110 (3M+HK) painted in bright RLM 70/02/65 camouflage. Example of frequent occurence of different camouflaged planes in the same unit during the same time.

(P. Jarrett via B. Ketley)

No positive information about Bf 110 prototype painting schemes is known. Most probably they were painted RLM 01 Silber (silver) or RLM 02 Grau (grey) overall. The latter seems more probable. It is not impossible that the Bf 110V1 and V2 were painted RLM 63 Grau (grey) overall.

It is more interesting, however, to describe the camouflage scheme for the Bf 110A and B aircraft. These were painted according to the RLM 70/71/65 scheme. Similar pattern was also applied to the Bf 110C–1s and C–2s used by the Luftwaffe in 1939–1940. The Bf 110s were among the very first Luftwaffe aircraft to be repainted in the RLM 70/02/65 scheme. Such camouflage was applied to the aircraft assigned to ZG 2 in the Spring of 1940. The machines used in combat during the Battles of France and Britain bore both camouflage patterns described above.

It seems that during the Battle of Britain some Bf 110Cs and D–0s of SKG 210 were painted RLM 74/75/02 on the upper surfaces and RLM 76 underneath; this scheme was widely introduced on Bf 110s in the Spring of 1941. Photos show that the Bf 110C and D aircraft used in Greece in April and May 1941 by ZG 26 'Horst Wessel' elements were painted RLM 74/75/02+ RLM 76. This scheme was practically the Luftwaffe standard until the end of the war.

The North African (tropical) and Luftdienst painting schemes are worth mentioning here. The aircraft used by the Luftwaffe in North Africa were painted in two types of camouflage. One was to paint upper surfaces with the RLM 79 Sand (sand) and the lower surfaces with the RLM 78 Himmelblau (sky blue). The other was a modification of the standard temperate scheme, and consisted in adding RLM 79 to the upper surface colours. The lower surfaces were painted RLM 76 Hellgrau (light grey) or overpainted RLM 78. Both schemes

The Messerschmitt Bf 110 B–1 in standard RLM 70/71/65 camouflage, back view. Clearly visible borders between colours.

(MVT via M. Krzyżan)

were used in parallel in Luftwaffe units in the same time.

The Luftdienst Bf 110s were sometimes painted in a two–tone scheme, with upper surfaces RLM 70 Dunkelgrün (dark green), and lower surfaces in RLM 65 Hellblau (light blue).

A complete description of Bf 110 camouflage patterns is to be found in the series of books about Luftwaffe camouflage and markings, published by the AJ–PRESS and available from Books International.

The Nachtjagd (night–fighting) Messerschmitt Bf 110s were painted differently from the day–fighting machines. The first scheme was to paint the aircraft matt RLM 22 Schwarz (black) overall.

In 1942 a new painting scheme was introduced, where the upper surfaces were painted in grey tones of RLM 74/75/76/77, and the lower surfaces in RLM 76. The guidance unit machines were sometimes painted RLM 77 Himmelgrau (sky grey) overall. The lower surfaces could be painted black, or RLM 76 Hellgrau (light grey). Sometimes both colours were used, e. g. underwing surface inboard of and including the engine nacelle was black, while the outer wing portion was light grey.

Also the colour patterns evolved in time. At first this was the classic Luftwaffe splinter pattern, and later it gave way to the spot pattern superimposed on the splinter scheme (on upper wing and tail surfaces). Another frequent style was the so called spiral camouflage introduced in 1943, which consisted of special sophisticated patterns made up of spiral lines in several colours, which had the effect of deforming the aircraft's outline.

The aircraft exported to Hungary, Romania and Italy bore standard German camouflage. The only difference lay in the national markings being applied in place of the Balkenkreuz insignia.

Another important element of the Bf 110 markings were the means of quick identification. The most often used was to paint the front of the engine nacelles in bright colours; white (Battle of Britain) or yellow (Balkans). Colour bands around the rear fuselage formed another quick identification element; such bands were applied on the Eastern Front (yellow) and in North Africa (white). The Bf 110Gs assigned to the Reich's air defence bore appropriate markings. In the blind flying training schools the aircraft carried twin wide (150–200 mm) yellow bands around the rear fuselage. Sometimes the forward fuselage was painted yellow or white.

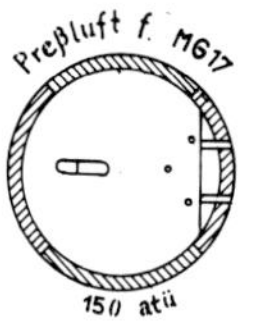

1 (Compressed air for MG 17) – inscriptions red, two part, two colour border; left and right part – red, upper and lower part – green

2 Brown triangle with white fringe, inside inscription Rötring (oil trade mark) – black

3 Fuel filler designation with its octane number (87) – yellow triangle with white fringe, numbers – black

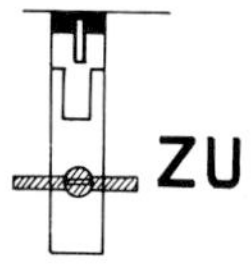

4 Engine cowling lock with red line of identation, ZU (close) letters – red

5 Name plate of propeller manufacturer placed on the front side of blades – black signs on the white background

6 (Oxygen for respirator) – circle with white-blue segments, inscriptions black

7 Manufacturer identification plate; black on the silver background

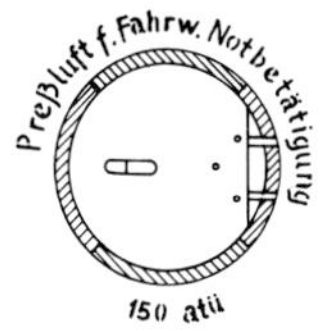

8 (Compressed air for emergency undercarriage lower) – black inscriptions; segment two colour fringe, left and right part – red, upper and lower part – green

9 Designation of the external power supply terminal – red circle, black letters

10 Designation of engine cooling liquid filling connector – white cover, brown fringe, black inscriptions

11 Designation of the first aid kit hatch – red cross on the white circle

Hier anheben
Vor aufbocken
m. 300 kg.
belasten

O

12 Designation of the tail lifting point – letters and circle around the lifting hole black

Einstegleiter Knopf drucken

13 (Access ladder, push button) – black inscription, only on the left side of fuselage

Reifendruck 4,5 atü

14 (Tires pressure) – black inscription

Nicht anfassen

15 (Do not catch) – red inscription

Hier
aufbocken

16 (Support here) – black inscription on the height of 18 frame from down

Hier
eingreifen

17 (Catch here) – black inscription over flap cover surrounded by black fringe

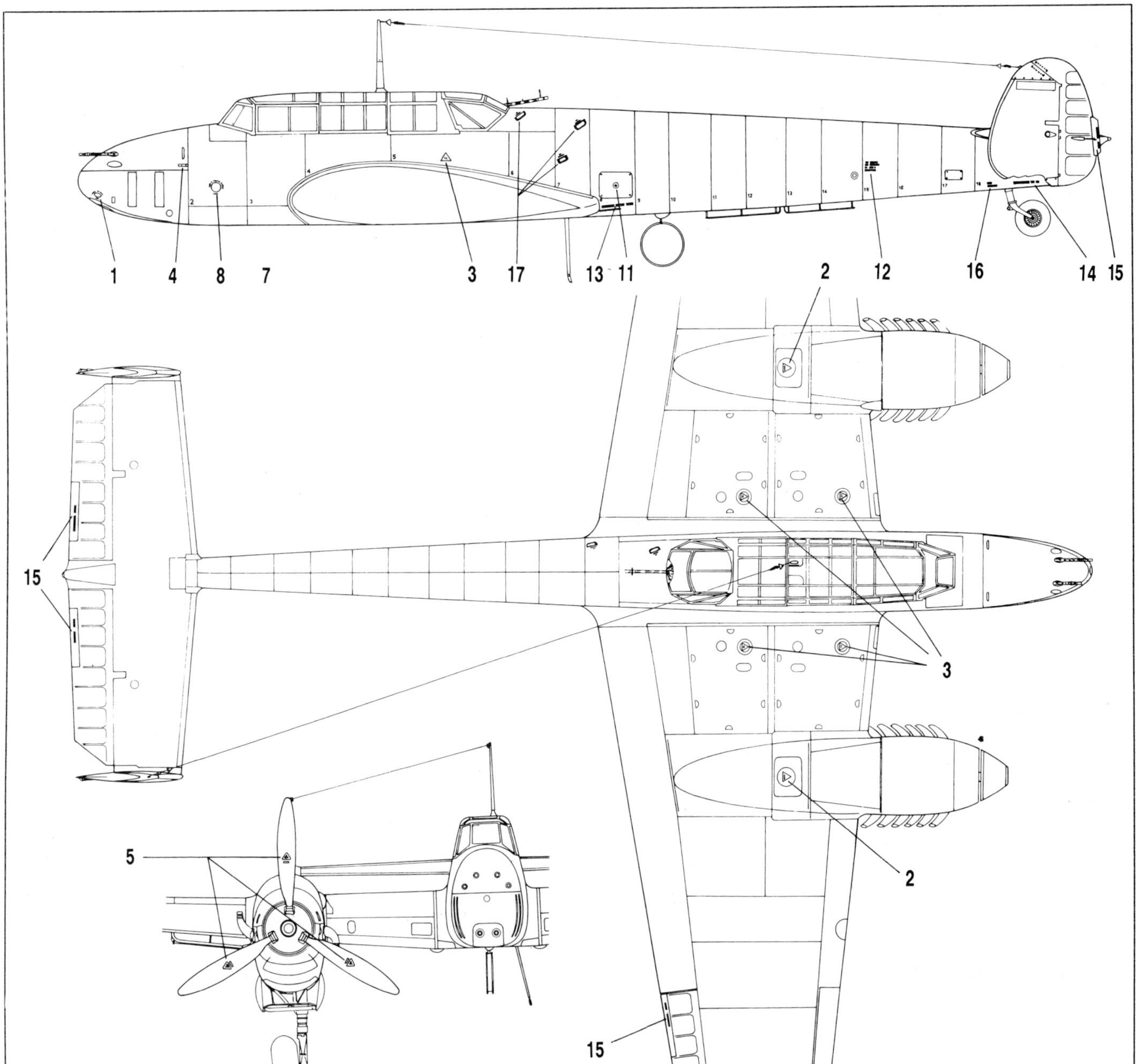
1 4 8 7 3 17 13 11 2 12 16 14 15
15
3
5
2
15
Designation of fastening position of ruder fixing plates – red dashed lines
Research and traced by Krzysztof M. Żurek
1/72 scale
Designation of frame number – black number just after plates joint
10 4 12 6 9 3
3
4
15
14 16
0 1 2 3 4 5

TECHNICAL DESCRIPTION OF MESSERSCHMITT Bf 110

The Messerschmitt Bf 110G–2 was a two–seat, twin–engined, all–metal cantilever low wing monoplane with retractable landing gear.

The semi–monocoque fuselage with oval cross – section was made up of two halves fixed together on top and at the bottom, along the vertical plane of symmetry.

The frames were of top–hat section, while the longerons were Z–bars. The fuselage skin was of flush riveted dural sheets. The wing centre section was integral with the fuselage; the nose section was fixed to the forward fuselage, and housed fixed gun and cannon armament. Further aft, the crew cockpit was placed with the pilot's seat forward and that of gunner/radio–operator aft. Forward of the pilot an instrument panel was mounted with a set of flight/navigation instruments and engine controls. The port console housed flap, trim tab, oil cooler, fuel pump, landing gear, magnetos, and propeller pitch controls. The starboard panel included starter rods, fuel system valves, rudder trim tab, and starting pump controls.

The pilot's seat was stamped from dural sheet, shaped to accomodate a seat–type parachute, and armoured with a back plate extending up behind the pilot's head. The aircraft controls were in form of a control stick and rudder pedals. The space behind the seat was occupied by radio equipment, ammunition magazines or an additional fuel tank of 540 litres capacity. The rear cockpit housed the gunner/radio–operator's seat. The cockpit cover was in five–pieces, the forward portion opening up– and sideways, and the gunner's cockpit cover opening to starboard. The aerial mast was fitted on top of the cockpit. Behind the cockpit, the fuselage housed more radio equipment, oxygen bottles and first aid kit. Cockpit entry was from the port side (with steps in the fuselage near the wing trailing edge). The forward windscreen was of 57 mm thick armoured glass. Standard equipment included a Siemens K4–1 autopilot.

The wing was of two–piece all–metal construction, of single–spar design with flush–riveted stressed skin. Between the fuselage and the engines the skin was reinforced with top–hat section longerons and stiffening elements. The wing structure consisted of latticed ribs spaced evenly along the span, and top–hat section longerons with stiffening Z–bar elements.

The ailerons were slotted and fitted with horn balances and trim tabs, and were fabric covered. Handley Page automatic slats were included in the wing leading edge. The wing trailing edge featured slotted metal–covered flaps. The port wing leading edge incorporated a 500 W landing light, and navigation lights were fitted at the wingtips. Each wing was attached to the fuselage at four points. A pitot tube was fitted to starboard wing.

The all–metal tail surfaces were of semi–monocoque design, with twin vertical fins and rudders. The horizontal stabilizer incidence was adjustable between +3.0° and –0.9°. The rudders and elevators were mass and aerodynamically balanced, fabric – covered and fitted with trim tabs. The horizontal stabilizer was adjusted by turning a knob in port side of the cockpit.

The engine-propeller-landing gear unit of Messerschmitt Bf 110 G–4.

(Fot. J. Wróbel)

The landing gear was in classic layout with a tail wheel. The wide track main gear with single legs hydraulically retracted backwards into the engine nacelles. Starting from the Bf 110F, the aircraft was equipped with an emergency pneumatic wheel lowering system fed from a container with air pressure of some 8 MPa. Wheel bays were covered with twin doors. The main wheels were fitted with double drum hydraulically operated brakes (starting from the Bf 110F), fed from an autonomous circuit, independent of the main hydraulic system. VDM type main wheel legs with oleo shock–absorbers carried 935x345 mm main wheel tyres. The earlier versions featured 815x320 mm tyres (Bf 110B) or 875x320 mm (Bf 110C and D) with single brakes. The tail wheel was fitted with a 465x165 mm tyre (starting from the Bf 110D), on a VDM fork, with sprung shock absorber and (starting from Bf 110F) an oleo damper. The Bf 110B and C featured a retractable tail wheel, but this was usually fixed in the open position by the ground crew in units. The Bf 110B had a 350x135 mm tail wheel tyre, while the Bf 110C – 380x150 mm. The main wheel tyre pressure was O,50 MPa, the tail wheel pressure – 0,325 MPa. In the earlier versions the tyre pressure was higher – 0,65 MPa in main wheels and 0,48 MPa in the tail wheel. The wheel bands were fitted with an electric system to indicate low tyre pressure.

The power plant (Bf 110G–2). Two Daimler Benz DB 605B–1 twelve–cylinder, in–line, inver-

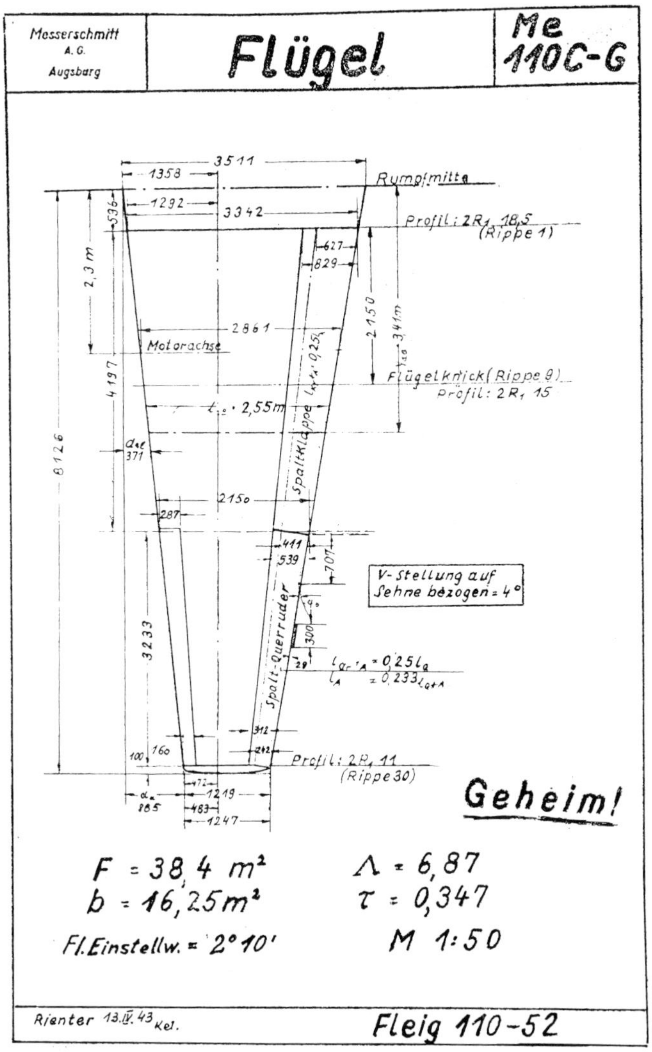

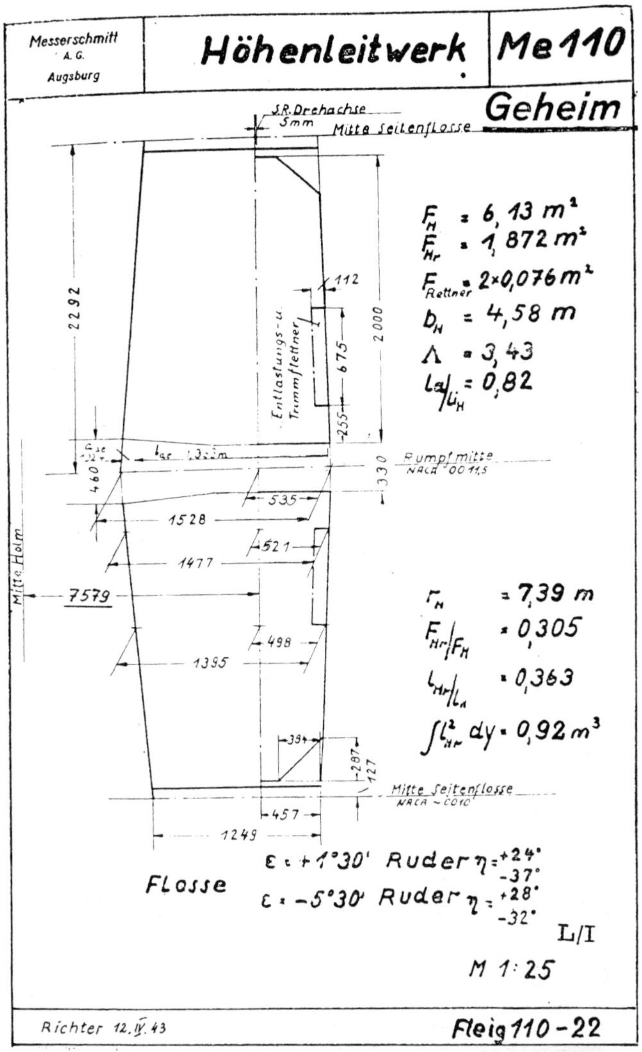

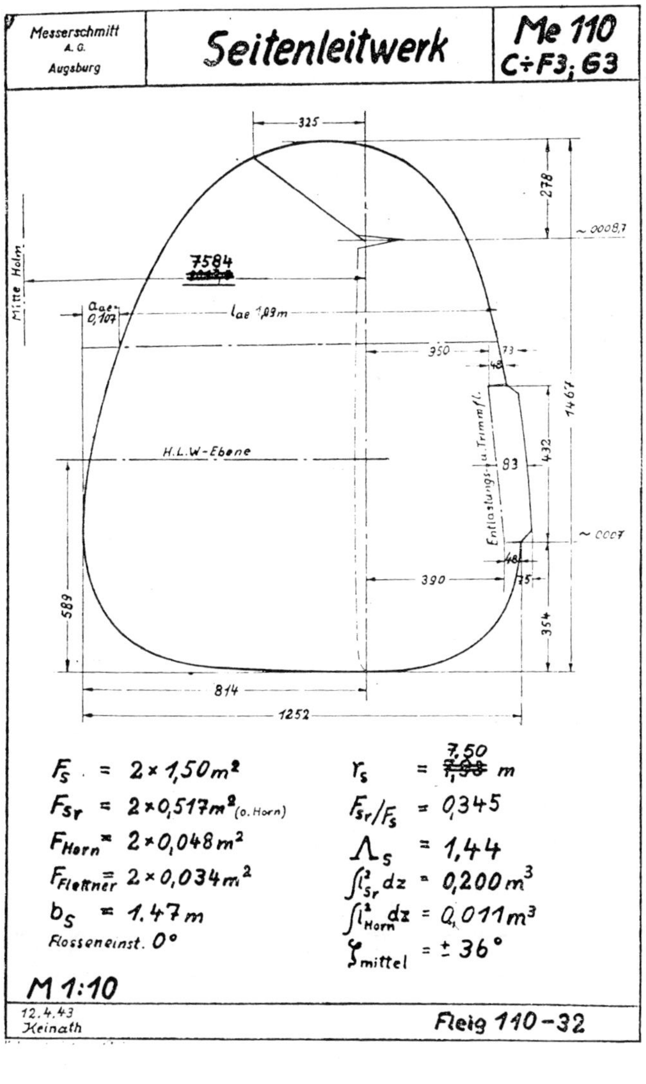

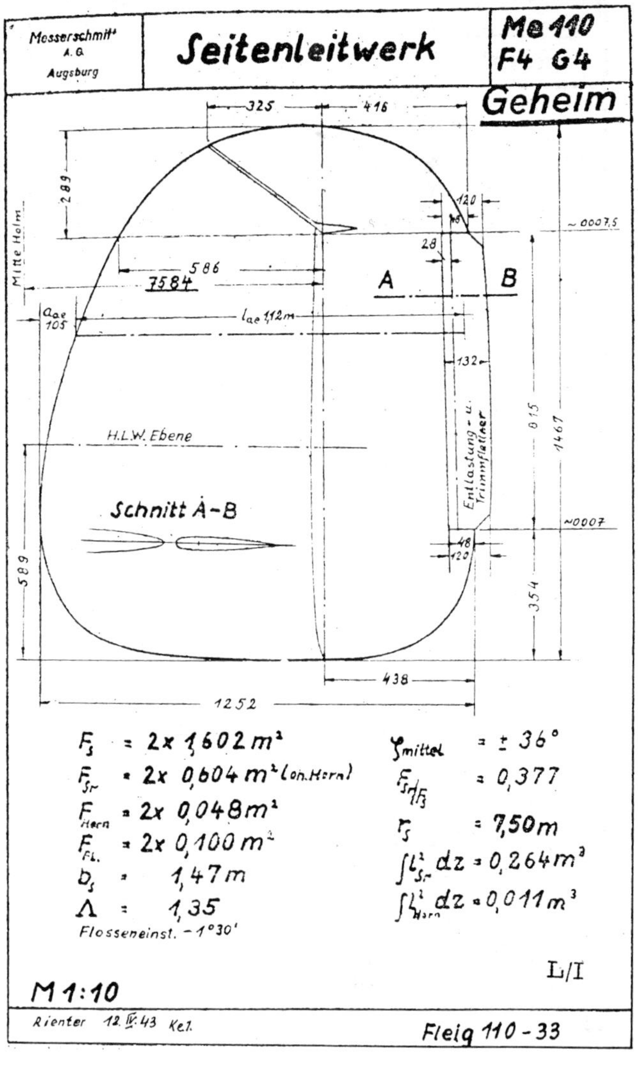

Factory drawings of airplane geometry (via M. Krzyżan)

Daimler Benz	Motorenmuster DB 601 A-B Kraftstoff B 4	Blattzahl: 1 Blatt: 1

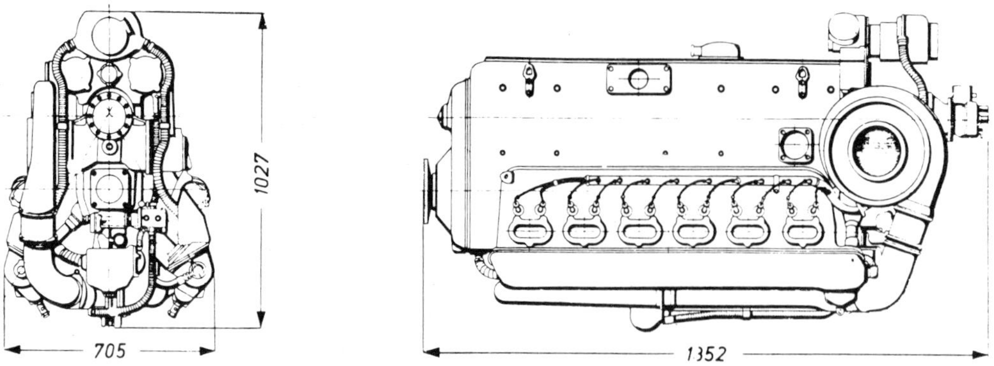

Höhe km	Leistungsstufe	PS	U/min	Ladedruck ata	Kraftstoffverbrauch g/PSh	l/h
0	1' erhöhte Kurzleistung	1100	2400	1,40	250 + 20 g/PSh	375
0	5' Kurzleistung	990	2400	1,30	220 + 12 g/PSh	300
0	30' erhöhte Dauerlstg.	910	2300	1,23	220 + 12 g/PSh	275
0	Dauerleistung	810	2200	1,15	220 + 12 g/PSh	245
4,5	5' Kurzleistung	1020	2400	1,30	220 + 12 g/PSh	310
5,0	30' erhöhte Dauerlstg.	960	2400	1,23	220 + 12 g/PSh	300
4,9	Dauerleistung	860	2200	1,15	210 + 12 g/PSh	255
5,5	Sparsame Dauerlstg.	800	2200	1,10	210 + 12 g/PSh	240
10,0	Notleistung	500	2400	0,70	300 + 12 g/PSh	210

Daimler Benz	Motorenmuster DB 601 E-G Kraftstoff B 4	Blattzahl: 1 Blatt: 1

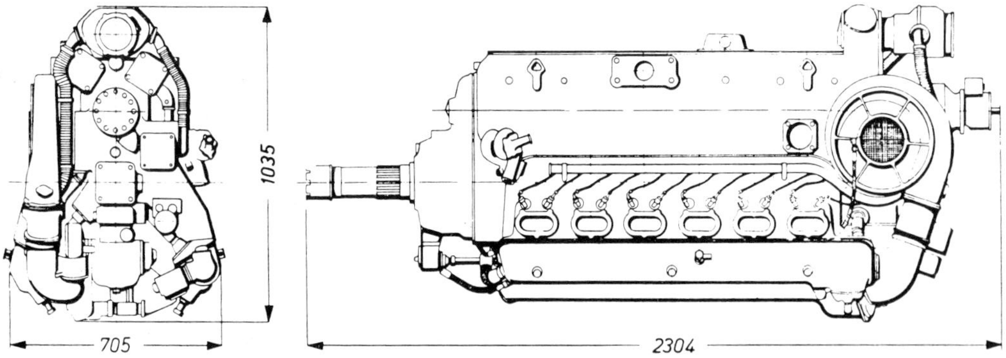

Höhe km	Leistungsstufe	PS	U/min	Ladedruck ata	Kraftstoffverbrauch g/PSh	l/h
0	Start- und Notleistung	1350	2700	1,42	235 + 10 g/PSh	445
0	Steig- und Kampfleistung	1200	2500	1,30	220 + 10 "	372
0	Höchstzul. Dauerleistung	1000	2300	1,15	210 + 10 "	290
4,8	Notleistung	1320	2700	1,42	235 + 10 "	440
4,9	Steig- und Kampfleistung	1200	2500	1,30	220 + 10 "	372
5,1	Höchstzul. Dauerleistung	1040	2300	1,15	215 + 10 "	300
5,1	Höchste Dauersparleistung	840	2000	1,0~~0~~5	205 + 10 "	240
10,0	Notleistung	700	2700	0,80	285 + 10 "	280

DB 601A engine, used in Bf 110C and D versions

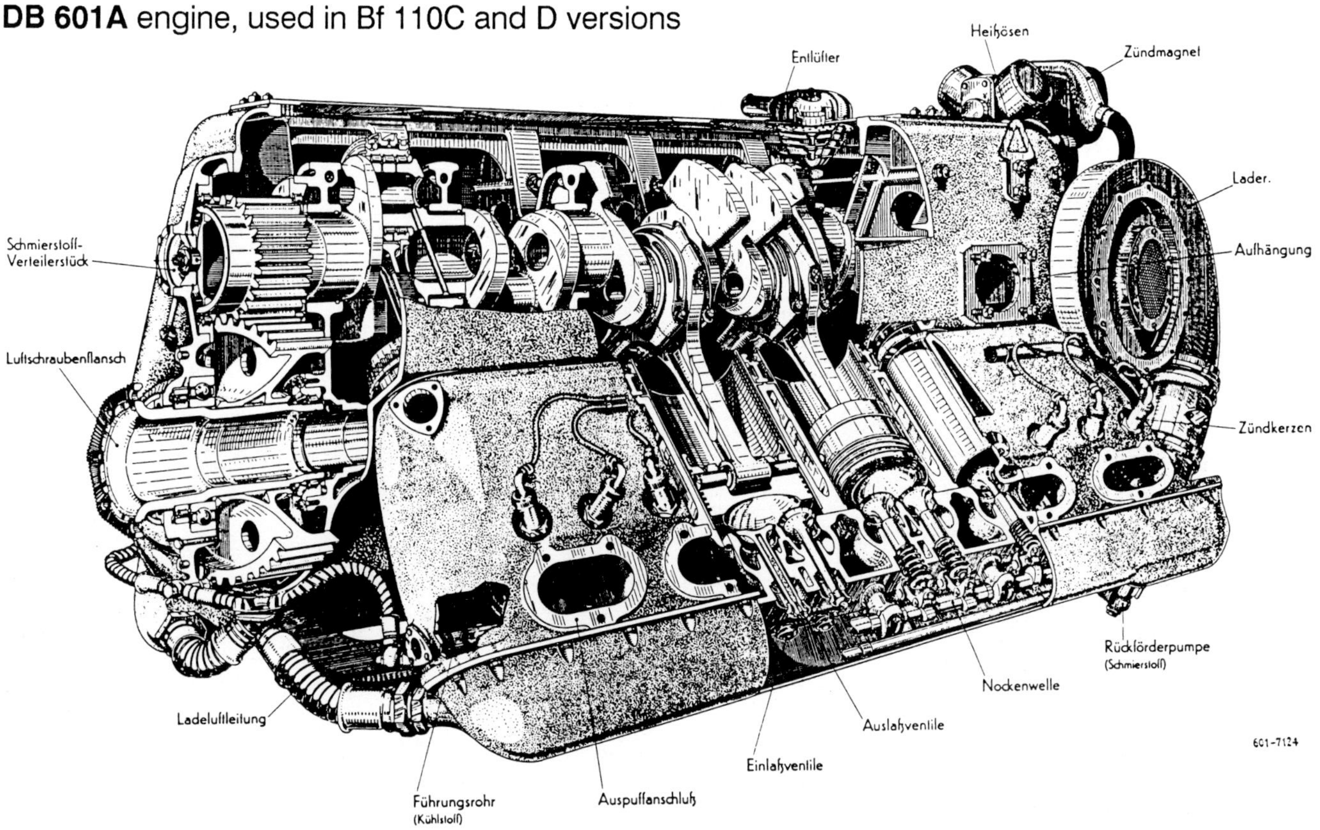

DB 601E–G engine, used in Bf 110F version

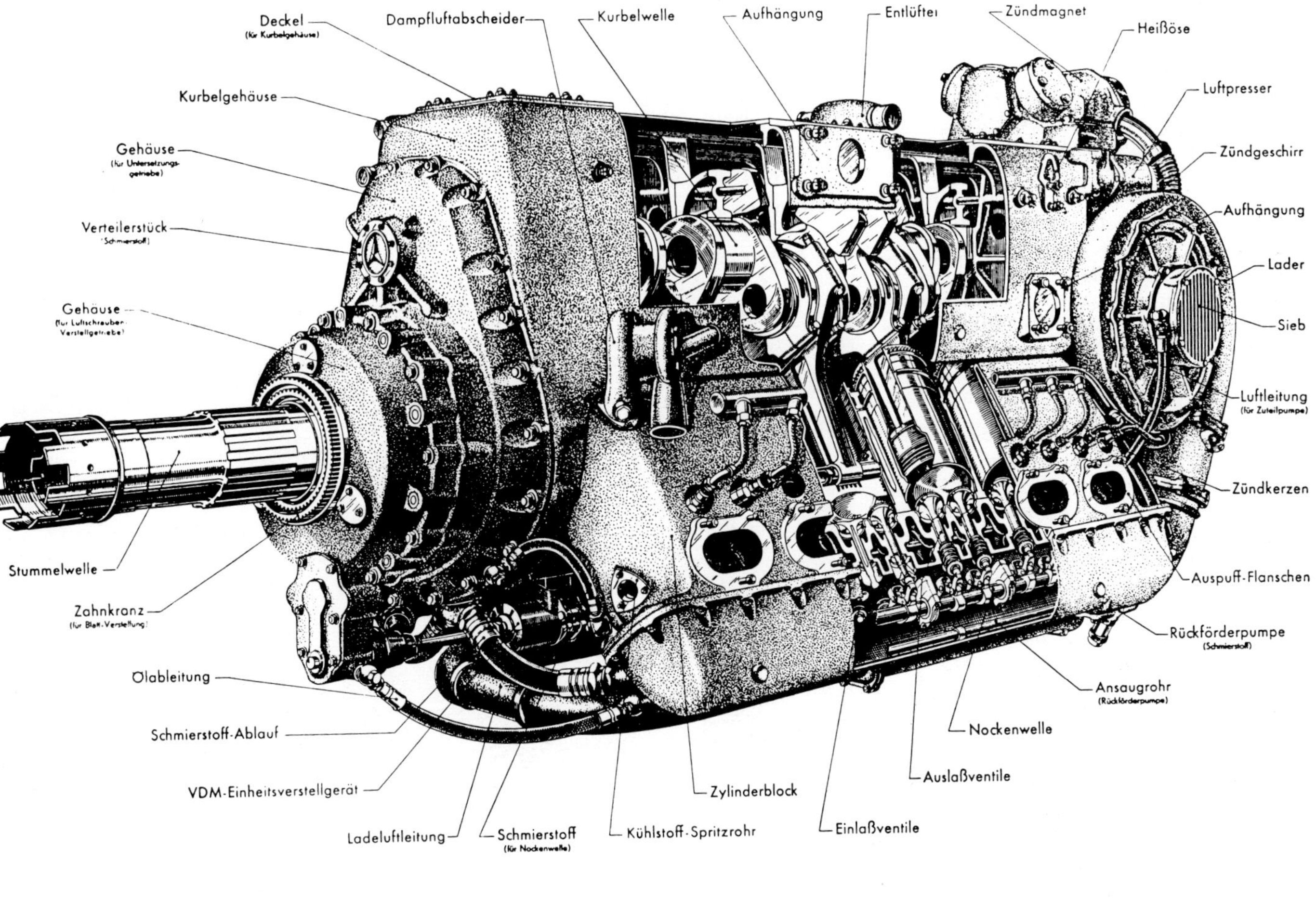

Main landing gear leg and part of engine cowling with oil radiator of Bf 110 G–4. *(J. Wróbel)*

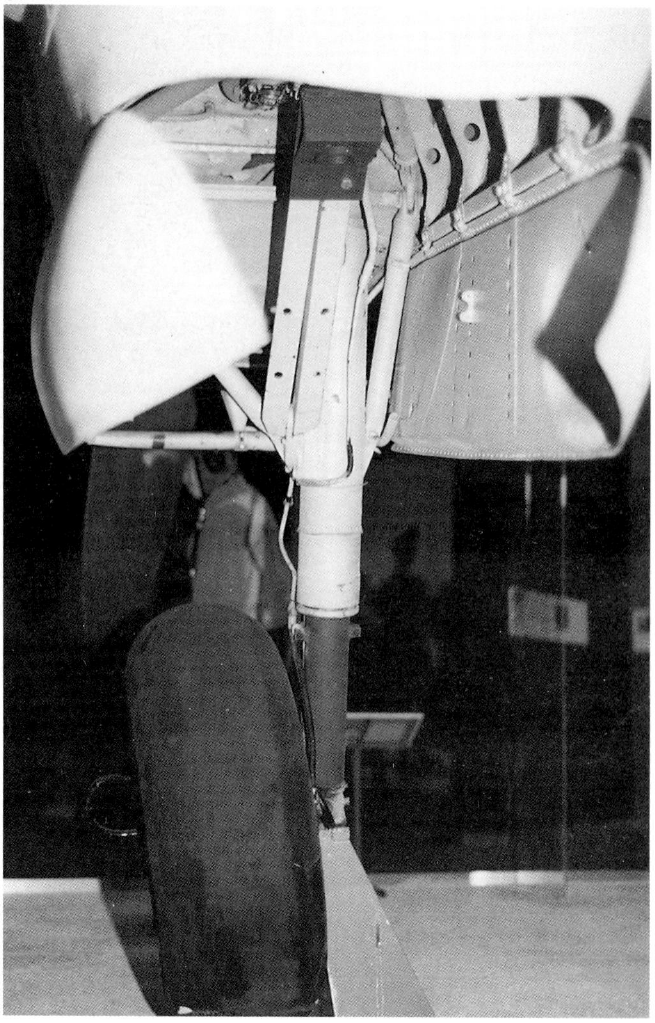

Main landing gear leg of Bf 110 G–4, rear view. Clearly visible part of wheelhouse. *(J. Wróbel)*

ted Vee engines, rated at 1,085 kW (1,475 hp) for take–off and 960 kW (1,310 hp) sustained power. An engine weighted 728 kg. The engine was driving a VDM 9–12098A/B constant–speed propeller of 3,375 mm diametre. The engine was cooled with a 1:1 water:ethylene glycol mixture with 1.5% addition of Schutzöl 39 anti–corrosion agent, in a separate cooling system. Each engine's cooling system incorporated a surge tank and underwing radiator. The airflow through the radiators was controlled with electric–operated flaps at radiator outlet. These were controlled with switches of the cockpit instrument panel.

Fuel system. Two selfsealing 375 litre tanks in the wings between the fuselage and engines in front of the spar, and two 250 litre tanks behind the spar. The fuel was B4 87–octane leaded petrol. For startup of a cold engine injection of startup fuel (gasoline plus ether) was used, the fuel fed from a special tank which, together with a hand–pump, was placed in the starboard panel in the cockpit. The aircraft could carry two underwing Junkers–type 300 litre jettisonable fuel tanks, which fed the forward wing tanks. The engine was fed via a circuit of pipes with valves and a filter set, the fuel pump was assisted by an electric pump fitted in each tank. The forward engine nacelle frame (Rib 6b) was a firewall, behind which a fuel filter set was fitted with emergency fuel cut–out valves.

The oil system consisted of two all–metal tanks of 50 litres capacity (filled up to only 35 litres, fitted in the rear engine nacelles) and SKF Fö 812 oil coolers fitted in the lower engine cowling, and a gear pump with oil thrower. The airflow through the coolers was automatic controlled by a thermostat, with movable flaps at the lower outlet. Intava–Rotring oil was used. The Messerschmitt Bf 110G–2 also featured an additional 75 litre jettisonable oil tank under the fuselage.

The hydraulic system was used for lowering and retracting the landing gear and landing flaps.

Additional droppable fuel tank of 300 litres capacity mounted under wing of Bf 110 G. Clearly visible way of tank fastening. Over the tank visible engine cooling liquid radiator and flap in down position. *(J. Wróbel)*

Filling with ammunition strip 7.9 mm MG 17 machine guns. Clearly visible gun spindles with adjusting screws.

The hydraulic pump was driven by the starboard engine. The surge liquid tank was placed in the starboard wing above the wing nacelle. In case of emergency the flaps and landing gear could be lowered pneumatically.

The oxygen system consisted of nine 2 litre spherical tanks, in three sets of three reservoirs, each with a separate valve and reducer. The set was fitted in the fuselage immediately behind the cockpit. It was filled via a common valve in the fuselage. The system pressure was 1.5 MPa. It was possible to install a second similar set in the fuselage, used in some Bf 110G–4s. The oxygen was fed to the respirators of the pilot (on the starboard panel) and the gunner (also on the starboard panel). The mixture ratio was automatically controlled depending on the altitude. Above 8,000 m pure oxygen was fed.

The electrical system was single–wire, screened, rated at 24 V nominal voltage. It was fed by two Robert Bosch LK 2000/24 generators (previously RKJ 2000–137) rated at 2,000 W, driven by the aircraft engines, and a 20 Ah battery in the rear fuselage. The master switchboard was in the pilot's cockpit, on the console. All circuits were protected with automatic fuses.

Radio equipment. The Messerschmitt Bf 110G–2 was equipped with the FuG 10P set of radio transmitters–receivers consisting of S.10K and S.10L transmitters, and E.10K and EZG receivers, plus the APZ 6 radionav aid. These used the aerial suspended between the mast on top of the cockpit and the top of fin, and a wire aerial dragged under the fuselage. A circular PRE 6 aerial was used for radio–navigation. The crew used an EiV intercomm system which used the FuG 10. Some aircraft were equipped with the FuG 25a IFF fitted with a self–destruction unit. The FuG 25a used a mast aerial under the fuselage. The FuBl 2 radionav unit was used for blind landing, and this used a two–dipole aerial placed flat under the rear fuselage.

The electronic system consisted of a Peil G V radiofinder, FuG 101a radioaltimeter and PFK–13 compass.

Equipment: The Messerschmitt Bf 110G–2 aircraft could be fitted with a tropical system comprising air filters (made by the Italian Marinelli company) on the engine intakes, dust–proof sealing, hand firearms and water/food kit. In arctic conditions a winter set was used, with skis, food, and a heater. A 27 mm Walther LP flare pistol was also fitted in the aircraft with the

The Rüstsatz 3 (R3) consisted of pair of 30 mm MK 108 cannons mounted in the later versions of Bf 110. On the photographs clearly visible grooved cartridge chamber cover. Also was used cannons with smooth cover of this chamber (see p. 56).

(MVT via M. Krzyżan)

Overview of the MG 17 installation

a. MG 17 machine guns
b. Cast mount
c. Steel mount
d. Feed chutes
e. Ejection chutes
f. Ammunition boxes
g. Compressed air bottles
h. ESK 2000b gun camera
i. Colimation tube
k. KVK 17
l. SVK 42B
m. A-firing button
n. SKK 404-2
o. Revi C12/D gunsight
p. Circuit breaker on main circuit breaker panel

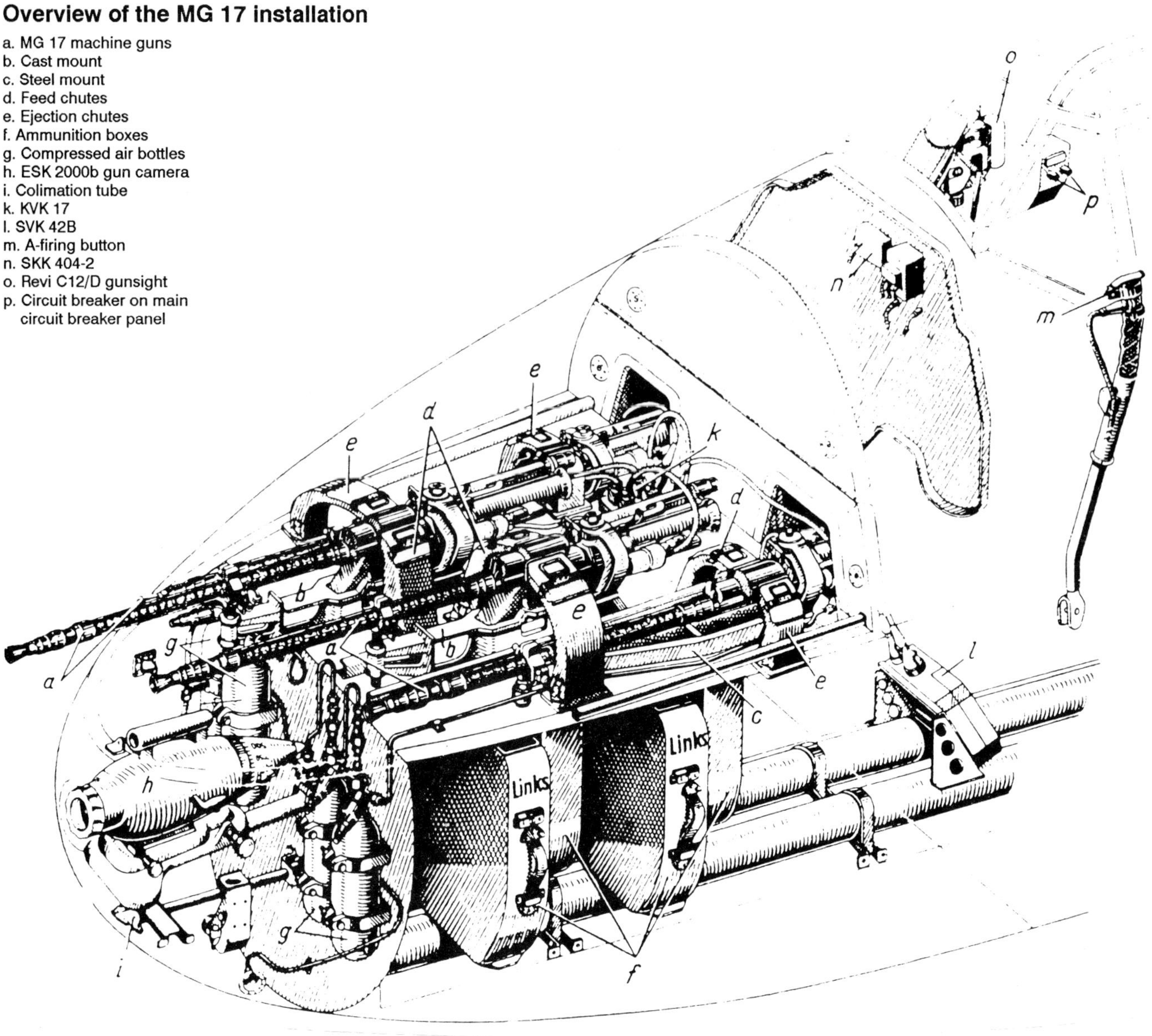

Overview of the fuselage MG 151/20 instalation

a. MG 151/20 cannon
b. Blast tubes
c. Ammunition boxes
d. Ammunition feed chutes
e. EDSK-B1 gun cocking control boxes
f. SVK 2-151/131E fuse and distributor box
g. Relay unit
h. Underfuselage fairing
i. SZKK 2 armament switch, rounds counter and control box
k. B-firing button
l. SKK 404-2
m. Circuit breaker on auxiliary circuit breaker panel

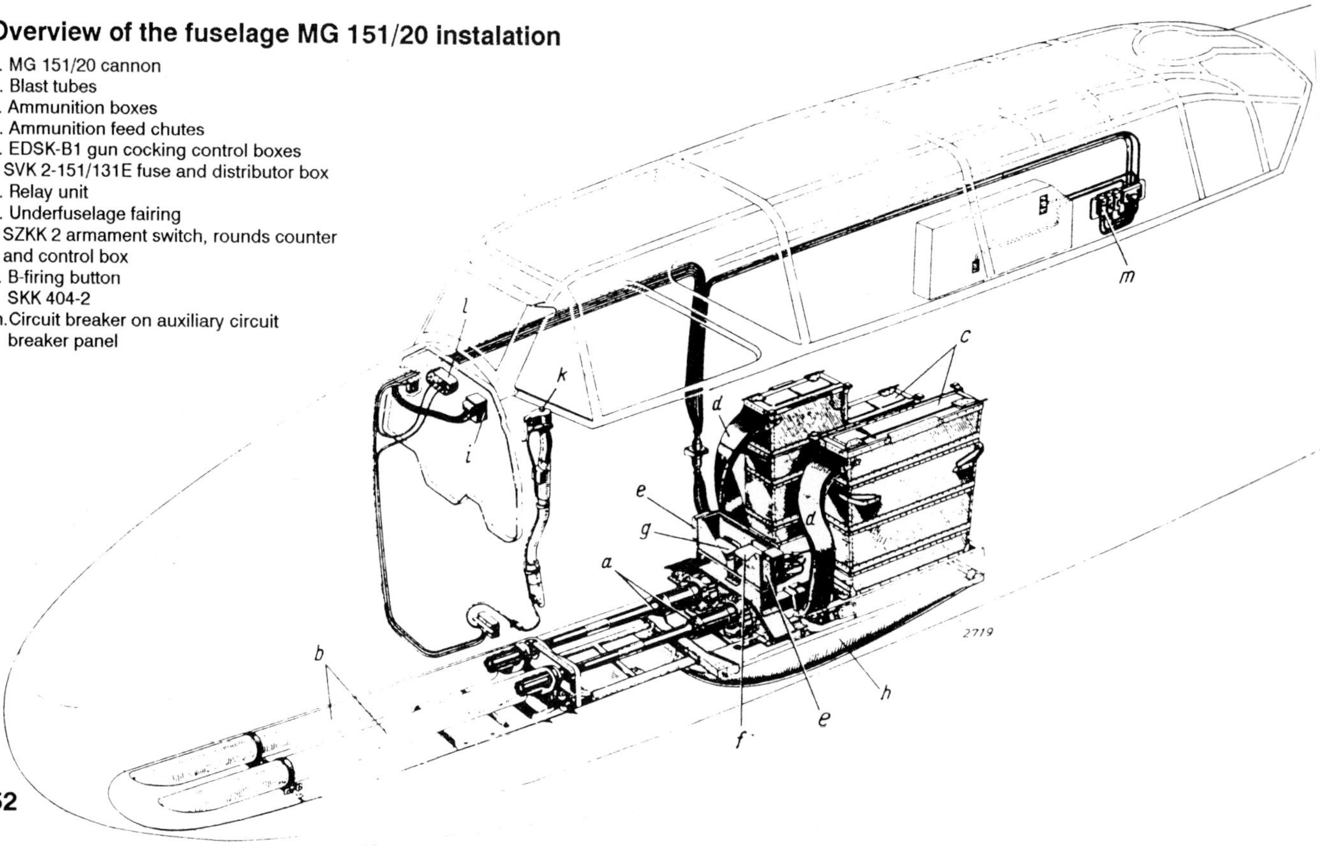

Rüstsatz 3 – overview of the MK 108 weapons instalation on Bf 110 G version

1. Left MK 108 cannon
2. Right MK 108 cannon
3. Empty cartridge ejection chute
4. Ammunition feed chute
5. Ammunition boxes
6. Compressed air bottles with DHAG 4 (later DHAG 5) pressure reduction unit
7. External connection
8. Valve for EPD (cocking unit)
9. Valve for EPA (firing unit)
10. Forward gun mount
11. Adjustment unit
12. Switch box panel
13. SZKK 3 switch, ammunition counter and control box for two cannons
14. Firing button
15. Blast tube
16. Barrel sealing units

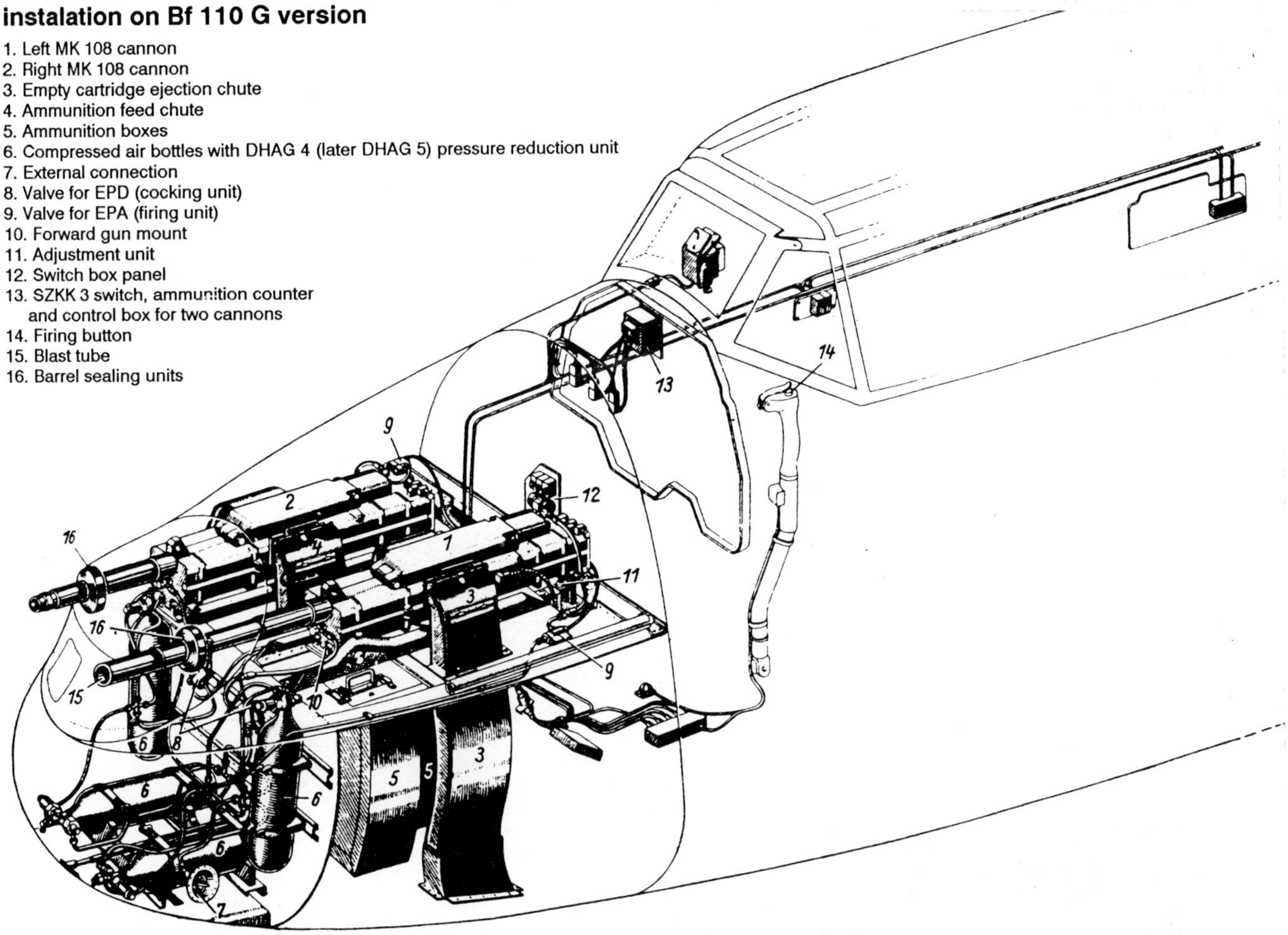

Overview of the reconnaissance camera installation

a. Aerial camera with film cassette
b. Camera mounting frame
c. Overlap control
d. Drive motor
e. Film advance control
f. Viewing window
g. Sliding panel
h. Camera opening
i. Forward fuselage panelling
k. Circuit breaker A150
m. Circuit breaker N16
n. Magnetic switch N17
o. Distributor N18
p. Blind mounting
q. Control light
r. Plug connector N6/N7
s. Przyłącze elektryczne N19/N20
t. Quick break switch N11

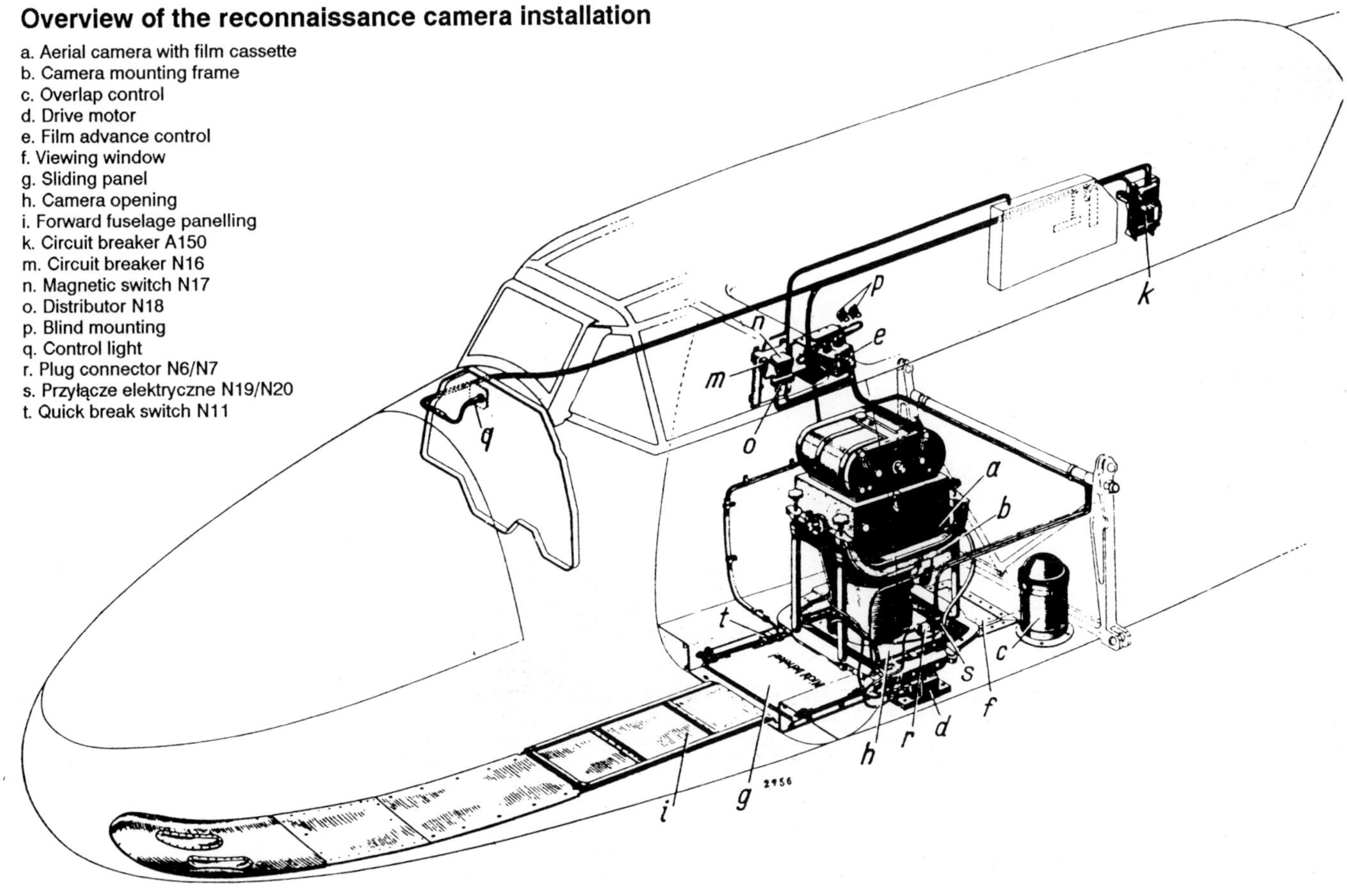

Rüstsatz R1 – overview of the 3,7 cm BK installation

a. BK 3,7 cm cannon
b. Cannon mounting
c. Weapons carrier
d. Left housing
e. Compressed air bottles
f. Compressed air triggering unit with EPD-FF/C cocking unit
g. Triggering unit
h. Brake cylinder
i. Ammunition clip carrier
k. Ammunition clip
l. Cable
m. Forward mount torsion bar
n. Locking lever
o. Rear mount torsion bar
p. Fairing
q. Forward fairing unit
r. Barrel protection cuff
s. Blast protection panel
t. Electrical connector cable
u. Plug connector
v. SZKK 2
w. SKK 404-2
x. Circuit breaker P2
y. B-firing button on KG 13 control grip

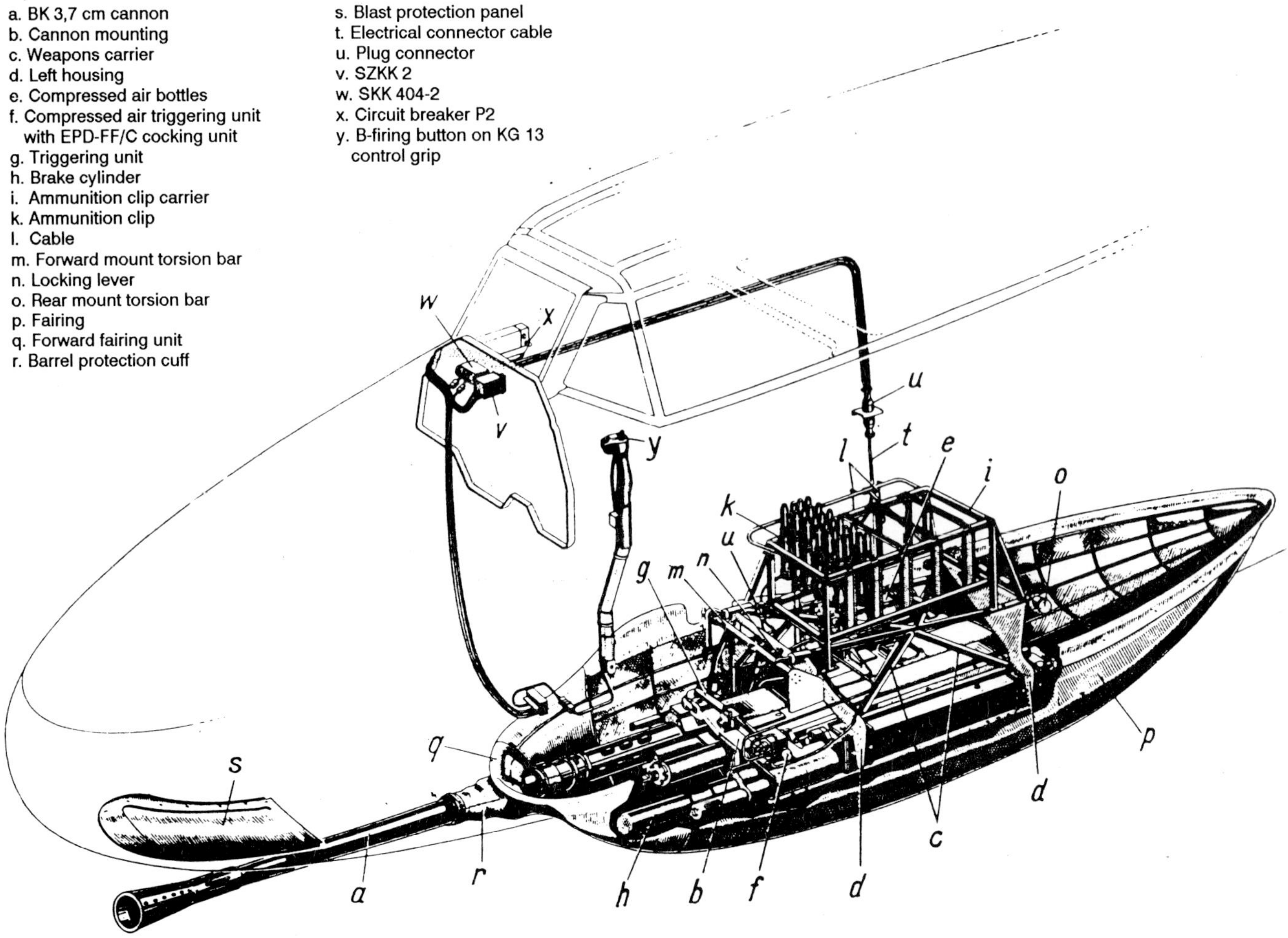

Rüstsatz M1 – overview of the two MG 151/20 cannon installation

a. Two MG 151/20 cannons
b. Ammunition boxes
c. Ammunition feed chutes
d. Ejection chutes
e. Weapons carrier
f. Fairing
g. Blind mounting
h. Plug connectors
i. Auxiliary circuit breaker panel
k. Circuit breaker
m. Indicator lights
n. SKK 404-2
o. B- firing button

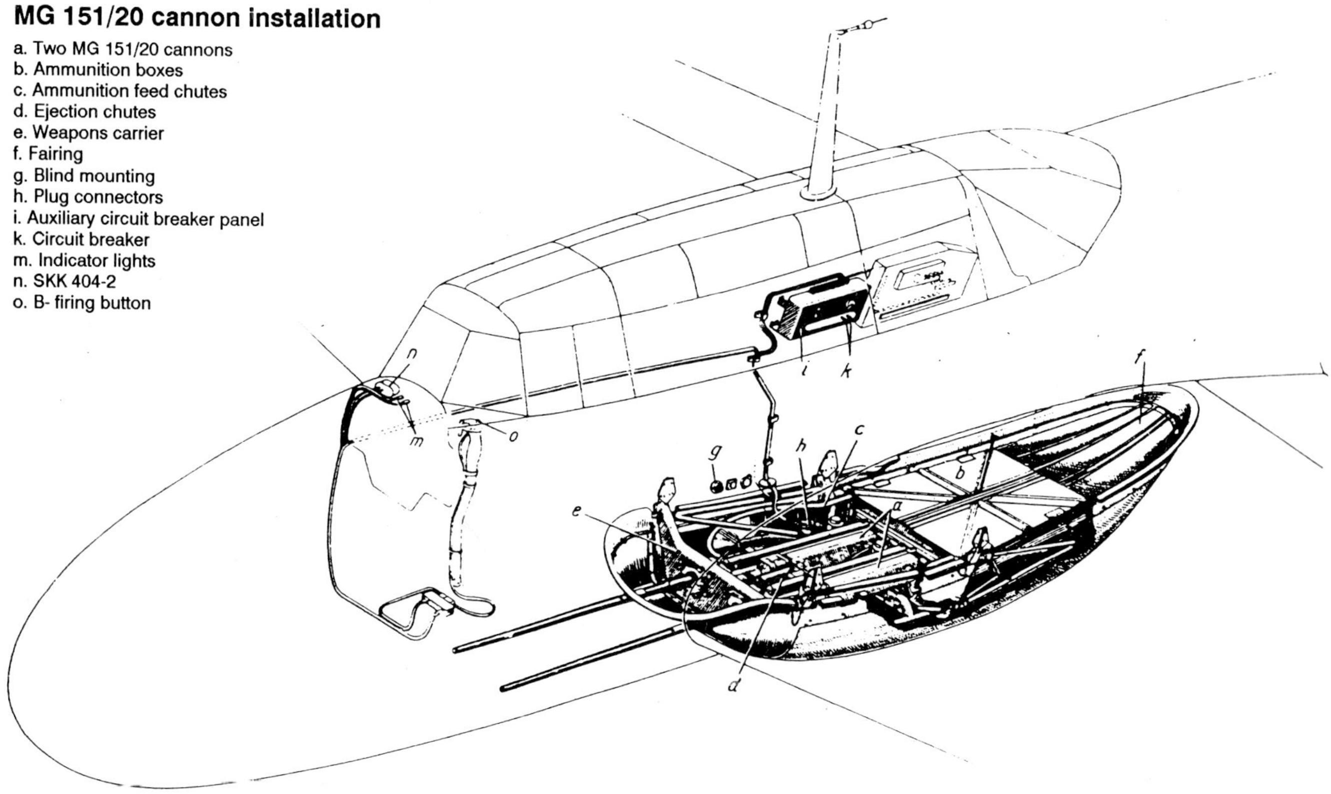

Rüstsatz 8 – overview of MG FF *Schräge Musik* installation

1. MG FF/M cannon
2. Ammunition drum
3. Reserve ammunition drum
4. Compressed air bottle with pressure reduction unit and cut-off valve
5. Cartridge casing holder
6. Cocking units
7. Weapons mount
8. Weapons bracing truss

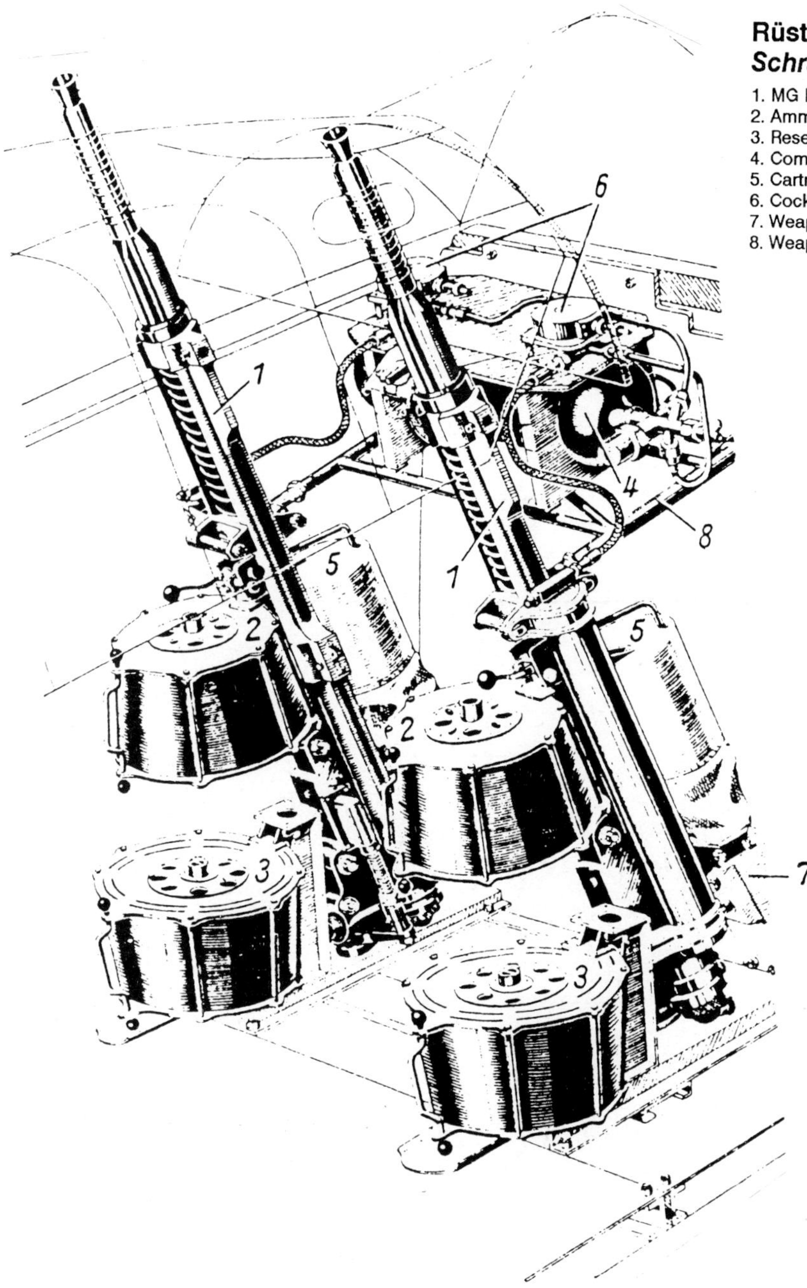

Rüstsatz R8 – overview of the upward firing armament system control

1. Circuit breakers P1 i P2
2. SZKK 3 armament switch, rounds counter and control box
3. Weapon loading (charging) button
4. Indicators
5. B-firing button
6. SVK 42
7. Main circuit breaker panel
8. EPD (cocking) unit
9. Distributor P100
10. Revi 16N gunsight
11. Distributor V1
12. Connector P355/P358
13. Gunsight sighting panel

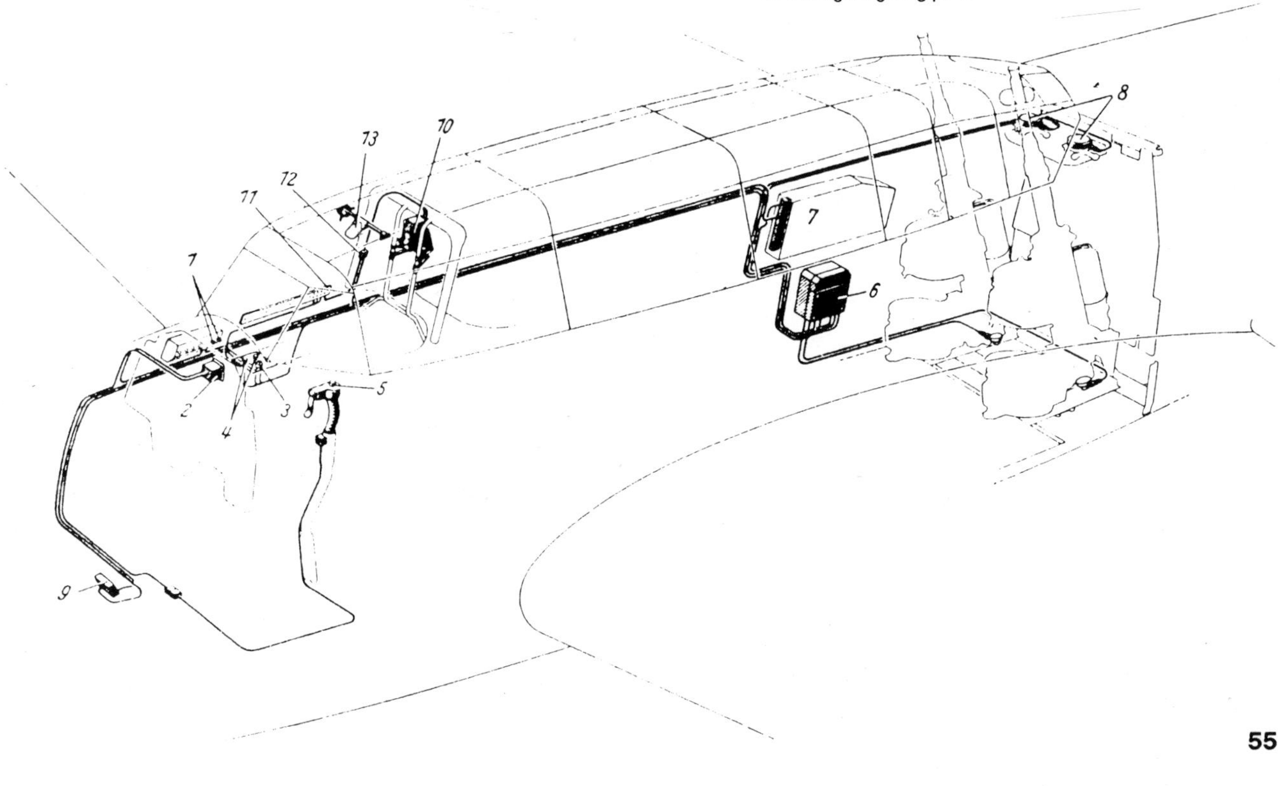

Another view of Rüstsatz 3 of Bf 110 G (pair of 30 mm MK 108 cannons) after removing of the armament cover. On the left photo one of the cannons was removed.

(MVT via M. Krzyżan)

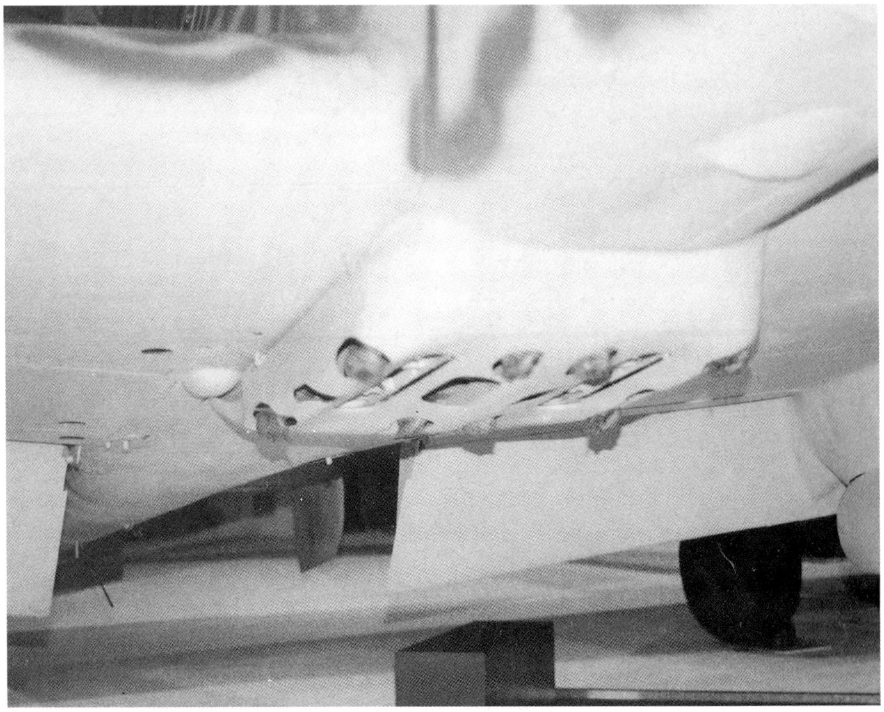

The ETC 500/IXb bomb racks for 500 kg bombs mounted under fuselage of Bf 110 G–4.

(J. Wróbel)

ability of shooting the flares outside in flight. A first–aid kit was also placed in the fuselage. When flying missions above the sea, the aircraft could take a single dinghy for both crew members, or two individual dinghies.

Armament. Normally the Messerschmitt Bf 109G–2 was armed with four Rheinmetall–Borsig MG 17 machine guns with 1,000 rounds per gun, fitted in the nose of the fuselage, and two 20 mm Rheinmetall–Borsig (Oerlikon) MG FF/M cannon with 180 rounds per cannon in three ammunition drums, or two 20 mm Mauser MG 151/20 cannon with 400 rounds per cannon for the port cannon and 350 rounds for the starboard one, fitted in the fuselage under the cockpit. The gunner had one twinned 7.9 mm Mauser MG81Z machine gun with 750 rounds. The bomb armament comprised two ETC 500/IXb bomb racks under the fuselage, enclosed with a streamlined fairing. Additionally a pair of ETC 50/VIIId racks could be fitted under each wing. The armament could vary, by way of field modifications and according to operational requirements.

MESSERSCHMITT Bf 110 SPECIFICATION

	V1-V3 prototypes	Bf 110 A–01 - 04	Bf 110 B–1	Bf 110 C–1	Bf 110 D–1	Bf 110 E–1	Bf 110 F–2	Bf 110 G–4c/R3
Dimensions: Span Length Height Wing area	16,80 m 12.50 m 4,10 m 45,6 m^2	16,80 m 12.50 m 4,10 m 45,6 m^2	16,80 m 12.523 m 4,10 m 45,6 m^2	16,20 m 12,07 m 4,12 m 38,4 m^2	16,20 m 12,07 m 4,12 m 38,4 m^2	16,20 m 12,07 m 4,12 m 38,4 m^2	16,20 m 12,07 m 4,12 m 38,4 m^2	16,20 m 12,07 m (13,05 m[f] 4,12 m 38,4 m^2
Weights: Empty Max take-off	– 5005 kg	– 5005 kg	– 5653 kg	4889 kg 6028 kg	– 7230 kg	– 6925 kg	5605 kg 7206 kg	5093 kg 9390 kg
Engines: Type Take-off rating	DB 600 A–1 986 hp	Jumo 210Da 680 hp	Jumo 210Ga 700 hp	DB 601 A–1[a] 1050 hp	DB 601 A–1 1050 hp	DB 601 N 1200 hp	DB 601 F 1350 hp	DB 605 B–1 1475 hp
Performance: max speed at height Landing speed Service ceiling Range Max range	502 km/h/3280 m – – – –	435 km/h/4100 m – – – –	453 km/h/4000 m – 7954 m 1715 km –	538 km/h/6000 m 150 km/h 10.000 m 1400 km 2600 km	538 km/h/6000 m 150 km/h 10.000 m 1400 km –	548 km/h/ – – 10.000 m 1400 km –	563 km/h/5360 m – 10.840 m 1200 km –	547 km/h/6940 m – 7954 m 900 km 2090 km[g]
Armament	3 -13 mmMG	4- 7.9 mm MG 17 1 - 7.9 mmMG 15	4- 7.9 mm MG 17 1 - 20 mm MG FF cannon	4- 7.9 mm MG 17 1 - 20 mm MG FF cannon 2-250 kg of bombs[b] or 2-500 kg of bombs[c]	4- 7.9 mm MG 17 1 - 20 mm MG FF cannon 2-500 kg of bombs[d]	4- 7.9 mm MG 17 1 - 20 mm MG FF cannon 2-500 kg of bombs[c] 4-50 kg of bombs[e]	4- 7.9 mm MG 17 1 - 20 mm MG FF cannon 2-500 kg of bombs[c] 4-50 kg of bombs[e]	2 - 30 mm MK 108 cannons 2 - 20 mm MG 151/20 cannons 1-7.9 mm MG 81Z

Notes: **a** -variant C-4B had 1175 hp DB 601N engine; **b** - in the C-4B variant on the ETC 250 bomb racks; **c** - in the C-7 variant on the ETC 500 bomb racks; **d** - in the D-3 variant on the ETC 500 bomb racks; **e** - on the ETC 50 bomb racks, it was possible to underslinge on them 2 additional fuel tank 900 litres capacity each; **f** - with radar aerials; g - with 2 additional 900 litres capacity fuel tanks.